Dare to Repair, Replace & Renovate

Also by Julie Sussman and Stephanie Glakas-Tenet

Dare to Repair

Dare to Repair Your Car

Dare to Repair, Replace & Renovate

Do-It-Herself Projects to Make Your Home
More Comfortable, More Beautiful, and More Valuable!

Julie Sussman & Stephanie Glakas-Tenet

Illustrations by Gavin Glakas

HARPER

NEW YORK • LONDON • TORONTO • SYDNEY

J

HarperCollins books may be purchased for educational, business, or sales promotional use. For information, please write: Special Markets Department, HarperCollins Publishers, 10 East 53rd Street, New York, NY 10022.

FIRST EDITION

Library of Congress Cataloging-in-Publication Data
Sussman, Julie (Julie Ellen)
    Dare to repair, replace, and renovate : do-it-herself projects to make your home more comfortable, more beautiful, and more valuable! / Julie Sussman and Stephanie Glakas-Tenet ; illustrations by Gavin Glakas.—1st ed.
        p. cm.
    ISBN 978-0-06-134385-8
    1. Dwellings—Maintenance and repair—Amateurs' manuals.   2. Do-it-yourself work.   I. Glakas-Tenet, Stephanie.   II. Title.
TH4817.3.S974 2009
643'.7—dc22   2008052011

09 10 11 12 13 ID/RRD 10 9 8 7 6 5 4 3 2 1

We dedicate this book to our dear friend, Linda Fuller, and in memory of her husband, Millard. As the co-founders of Habitat for Humanity International and the Fuller Center for Housing, these great humanitarians worked to provide decent, affordable housing to millions of people around the world. We take comfort in knowing that Millard is now home in God's loving arms.

Contents

Walls 103

Storage 149

Home Safety 169

Outdoors 193

Acknowledgments

It's said that the third time's the charm, but as authors, we think that the first and second times have been pretty good, too. How lucky are we that we get to have Collins for a third time as publisher of our Dare to Repair books? We thank you for your continued support of the series.

It's also said that you should never mix business with pleasure, but the good people at Lowe's make that impossible. It has been truly enjoyable working with Chris Ahearn, Kelly Persons, Karen Cobb, and Abby Buford to broaden women's knowledge of home repairs and projects. And an additional big thank-you goes to Jennifer Wilson for helping us reach out to manufacturers.

Never go into business with family, we've heard. But it's a good thing we did because we've been blessed to have Gavin Glakas as our outstanding artist for two books. You're brilliant . . . in so many ways.

To Deneen Howell, our agent / attorney extraordinaire. We're so grateful to have you guiding us with your infinite wisdom and gentle spirit.

To Dottie Hanson, thank you for your attention to detail, your diligence to completing the project, and your devotion as a friend. You are the "perfect one."

Special thanks to Dianne Layer, Brenda Daye, Faye Diel, and Patty Arena of Bob's Big Boy for always providing us with hot coffee and warm wishes as we traveled through Delaware.

And last but not least to our dear friend Laurie Wilner, whose commitment to friends knows no boundaries. Your skill, brains, humor, and handiwork made it possible for us to not only get the book done, but also get it done in time.

Personal Acknowledgments from Julie Sussman

I never would have imagined that my father, Warren Johnson, would not be around for the publishing of this book. My dad, a bibliophile who actually read each book, took great pride and joy in me being an author. I miss his humor, his honesty, and his brilliance. Yet, his presence in heaven gives me great peace because he is constantly showing me that he's still protecting me . . . just from farther away.

The world would be a better place if everyone had my mother as their mom. I knew as a very young child that my mother, Helene Johnson, was someone truly special. She is a gentle soul who raised five children without raising her fists; who gives to those in need, including animals, without needing anything in return; and who lives each day with a grateful heart, making everyone grateful to be near her. She is my hero.

Ann Walker, Mary Coyle, Chad Johnson, and Amy Marney are the absolute best people to have as siblings. Not only are they wonderful in their own right, but they had the good sense to marry outstanding people: Rod Walker, Ed Coyle, Donna Johnson (née Mitiska), and Bob Marney. And on top of that, they went and raised the greatest children on the planet: Patrick and Erin Walker; Andy and Brian Coyle; Graham, Bryce, Brooks, and Grace Johnson; and Caroline, Jill, and John Marney. And a special thanks to Patrick, the first grandchild to marry, for bringing Erin Bohan into our family.

To Stephanie, my Sister of the Traveling Tool Belts: I've shared more laughs with you than anyone else. And because of your wit, good nature, sensibility, patience, kindness, courage, and smarts, there's no one else I'd rather share a business with, too.

To my children, Chad and Rebecca: I adore you. I cherish you. And my love for you is eternal.

And last but not least, to my Jerry. Each day I truly believe that it would be impossible to love you more. And then the next day my feelings are déjà vu all over again. I love you . . . more.

Personal Acknowledgments from Stephanie Glakas-Tenet

Without my extraordinary family, amazing friends, brilliant co-author, gifted artist nephew, remarkable assistant, and techie genius friend, the third in the Dare to Repair series wouldn't have *dared* happen.

During a long year of writing, my husband, George, and son, John Michael, my family and friends patiently and lovingly indulged me in ways that only blood and friendship can explain. Thank you for your endless loyalty, infinite encouragement, and priceless patience.

My co-author, Julie Sussman, the yin to my yang and the yang to my yin, is my wonderful and complete opposite. Without her, I would be perpetually off balance. Nobody has more vision, imagination, or wit than she. For seventeen years, she has cajoled three books to completion by the sheer power of her personality—a truly remarkable woman.

Gavin Glakas, a gifted and accomplished artist in his own right, provided the beautifully detailed drawings for this book (as well as for *Dare to Repair Your Car*). His gentle soul and ceaseless commitment thrived throughout many tedious moments. It's a privilege to showcase his brilliance on canvas (www.gavinglakas.com).

Dottie Hanson was the secret ingredient to completing this project. Her devotion to detail, her patience and relentless spirit no matter what time of day or night not only guaranteed success, but also became a source of strength, joy, and everlasting friendship.

Every team needs technical and computer support. No computer glitch, power failure, or mechanical impediment stopped Larry from ensuring we had the best technology always. He was on perpetual call and never failed to deliver.

Last but not least, Saint Jerry, Julie's husband and my brother-in-spirit, who over these many years has worn more hats than a hat rack, has earned his permanent and deserved halo.

Introduction

ome is where your story begins.

Most of you are familiar with the children's book *If You Give a Mouse a Cookie*. The premise of the tale is that if you give a mouse a cookie, he'll want some milk to go with it, and then he'll want a straw to put in the milk. And he'll continue to need one more thing after another. The same holds true for tackling a home project—once you start, you'll find another one you want to do, and then another, and the list never seems to end.

Take for example a friend of ours who decided to spruce up her master bath by replacing the mirror. The new mirror looked wonderful, but it made the vanity look outdated. So that had to go. And then she realized that if she ever wanted to replace the flooring, she needed to do it before she installed the new vanity. And wouldn't it be easier to put down flooring without the toilet? And so on, and so on.

We wrote *Dare to Repair, Replace & Renovate* because we know that projects and repairs are never scarce, but time, money, and reliable contractors are.

Over the past few years, we have had the privilege of meeting thousands of you across the country at book signings, speaking appearances, Dare to Repair workshops on military bases, and Habitat for Humanity Women Build events. We loved hearing your stories of repair successes, and your desire to learn more and do more. But we also heard your frustration at not knowing how to get started: What tools do you need? What materials should you buy? Do you have the right information? Is the project over your head?

Wow, that's a lot of questions. But, hey, we think we've answered all of them in this book, our third in the Dare to Repair series.

We provided a pictorial of the tools and materials you need for each project, as well as some ideas for greening your home. We also did all the research, so you won't have to. We contacted government agencies, institutes, and nonprofit organizations for the most up-to-date information. And we reached out to manufacturers for their most cost-effective, user-friendly products, which we made sure to pass on to you. We even went to our favorite home improvement retailer, Lowe's, and asked for its list of top projects that every woman wants to learn. And, as always, we did every repair and project ourselves, so we can say with conviction, "If *we* can do it, *you* can do it."

Each of you may have a different reason for needing to do a project yourself. It may be that you can't afford the cost of materials *and* labor. Or maybe you've had horrible luck finding a dependable contractor. Or maybe you just want to raise the bar for what you can accomplish. No matter the reason, the rewards are the same: you saved money and you learned a new skill.

Your home may be the biggest investment you have, but it's so much more than a monthly mortgage payment. It's where you place your head at night, and it's where you literally wake up and smell the coffee. It's the address you give to your friends, and it's the haven you can't wait to return to after a long trip.

It's where your story begins.

In our first book, *Dare to Repair*, we invited you to step into the world of basic home repairs to become more resourceful and to save money. In *Dare to Repair Your Car*, we encouraged you to take the driver's seat when it comes to maintaining your car to minimize repairs and to stay safe on the road. In *Dare to Repair, Replace & Renovate*, we're challenging you to take those home projects that are on your *wish list* and move them to your new *can-do* list.

You know you can do it. *We* know you can do it. So, just do it. We *dare* you.

Plumbing

Our parents' generation never uttered the word "outdated." If something in the house was functional, it stayed. How else can you explain so many pink toilets in bathrooms across the country?

Although we agree with our parents' practicality, we believe in the rejuvenating power of an updated kitchen and bathroom. So, in this section there are no repairs, just projects, with most done just for the "ahh" factor.

Something else our parents never said.

Replacing a Garbage Disposal

Robbie was one of those people who never, ever put anything into the garbage disposal that wasn't supposed to go in. Not even peels from *one* potato. So she was shocked when the plumber told her a dishrag was to blame for the demise of the disposal and even more stunned when she got a quote for $250 to have the appliance replaced. Robbie declined, saying she'd already thrown in the towel once and had no intention of doing it again.

It was that last avocado pit, wasn't it? You just had to put it down the disposal. Now you're wishing that you had planted it, right?

The average life of a garbage disposal is eight to ten years. Of course, that time span can be greatly shortened by an uninvited object. Some appliances can be left broken for as long as your patience holds out, but a broken garbage disposal will clog a sink, as well as a dishwasher. So, here's a way to save money and get your kitchen up and running again.

Buying a New Disposal

When you purchase a new garbage disposal, it's best to stick with the same make and model as the old one so that you don't have to replace the mounting hardware.

Also, you want to make sure that the horsepower is the same, because a disposal with more horsepower could create an electrical outage. Garbage disposals are rated by either ½ or 1 horsepower. A disposal with ½ horsepower is the most common household type.

If you were thinking of replacing the kitchen faucet, now's the time to do it because the p-trap and garbage disposal will be removed, thereby giving you easy access under the sink. And since you'll be removing the p-trap, you may want to buy a new one or just new washers.

Getting Started

There are a lot of steps for this project, but they're all easy. Just take a deep breath and think about all the money you'll be saving!

Take the new garbage disposal out of the box and remove the contents that are inside (kind of like removing the giblets from a turkey, right?). Check that all the parts are included and read the manufacturer's instructions. Look at the illustrations and locate the parts on the disposal. This will definitely save you time . . . and grief.

Place a folded towel on the floor in front of the sink so you can kneel on it. Put a stand-alone flashlight near the sink to provide light—even if you're doing this during the day.

Remove everything from underneath the sink. Use the flashlight to get a good look at the garbage disposal, hoses, water lines, and p-trap.

Note: The following instructions are for replacing a garbage disposal that's hooked up to a dishwasher and is of the same make and model as the original disposal. If your new model differs from the old one, you need to follow the manufacturer's instructions for replacing the mounting.

Turning Off the Power

Turn off the power to the garbage disposal at the sink (the on/off switch) and have your helpful friend turn the power off at the main electrical service panel. Because your disposal is broken, it may be impossible for you to tell if the power is completely

Turning off the power

off; therefore, you need to test the on/off switch at the sink with a voltage tester. Remove the switch plate and touch the wires and terminals (screws on the sides of the switch) with the voltage tester. If the voltage tester beeps, then the power has not been shut off. Have your friend continue to try different circuit breakers or fuses until no beeps are emitted.

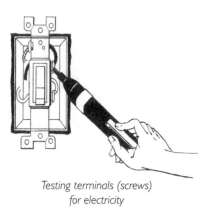

Testing terminals (screws) for electricity

Removing the Old Garbage Disposal

Removing the P-Trap

Place the small, shallow bucket under the p-trap. Using the plumber's wrench, turn the nuts counterclockwise and remove the p-trap and the extension pipe connected to the garbage disposal. If the nuts won't budge, spray WD-40 on them and wait a few minutes. Repeat.

The p-trap may stink, so be prepared to immediately wipe your hands with the wipes.

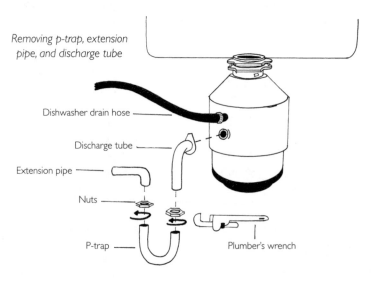

Removing p-trap, extension pipe, and discharge tube

Dishwasher drain hose

Discharge tube

Extension pipe

Nuts

P-trap

Plumber's wrench

Dump the contents of the p-trap and extension pipe into the bucket and rinse them out. Take a look at the washers on the p-trap and replace them if they're worn.

Disconnecting the Dishwasher Drain Hose

To remove the dishwasher drain hose from the garbage disposal, loosen the hose clamp and gently pull it out from the air gap. Do not remove the other end from the dishwasher inlet.

Removing the Discharge Tube

Remove the discharge tube by pulling it off the old garbage disposal. If it seems like it won't budge, just jiggle it around a bit and then give it a tug. Place it in the garbage can.

Be aware that a garbage disposal can be heavy, so be cautious when removing it.

Removing the Disposal

Insert the Phillips screwdriver or Allen wrench into one of the mounting lugs on the lower mounting ring. Push the handle of the screwdriver counterclockwise until the mounting assembly is loosened. Remove the garbage disposal.

Loosening disposal from mounting assembly

Disconnecting the Disposal from the Electrical Supply

Turn the disposal upside down and use the Phillips screwdriver to remove the square electrical cover plate. Loosen the green screw

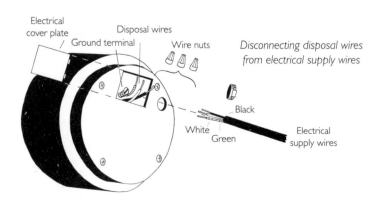

Electrical cover plate

Disposal wires

Ground terminal

Wire nuts

Disconnecting disposal wires from electrical supply wires

Black

White

Green

Electrical supply wires

(ground) with the Phillips screwdriver and remove the wire and wire nut. Disconnect the wires from the disposal from the house's electrical wires. Use the Phillips screwdriver to loosen the screw(s) on the electrical clamp connector to remove the wires from the disposal. Remove the electrical clamp connector and save.

Preparing the New Garbage Disposal

Because you're connecting the disposal to the dishwasher, you'll need to lightly tap out the knockout plug. Place the disposal on its side and insert the screwdriver into the dishwasher inlet, then hit the handle of the screwdriver with the hammer until you punch out the plug. You'll have to put your hand inside to actually remove it. DO NOT continue until you've found the plug.

Tapping out knockout plug

Connecting the Disposal to the Electrical Supply

Use the Phillips screwdriver to loosen the screws on the electrical plate, and remove the plate. Pull out the black and white electrical wires, but do NOT remove the cardboard insulation shield.

You'll need about ½ inch of exposed wire, so use the wire strip-

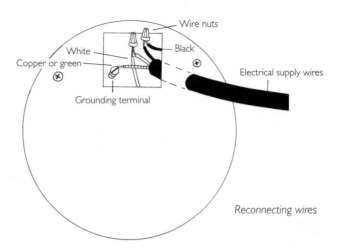

Reconnecting wires

per, if necessary, by placing the end of the wire into the proper hole of the wire stripper and the proper length that you want stripped. Clamp down and pull off the insulation. Slide the electrical cable through the access hole and secure it with the clamp connector.

Wrap the grounded wire around the green screw and tighten with a Phillips screwdriver.

Connect your home's white electrical wire to the white wire from the disposal with a wire nut twisted on top and connect your home's black electrical wire to the black wire from the disposal with a wire nut twisted on top. Finger-tighten the nuts.

Secure the electrical cover plate over the opening.

You're almost done!

Attaching the New Garbage Disposal

Have the Phillips screwdriver within easy reach.

Pull up on the mounting ring (above the ridges) and position the disposal so that the three mounting tabs are positioned to slide over the mounting tracks.

Push the disposal into the mounting assembly. Put the screwdriver into one of the lower mounting rings pulling it toward the right to tighten. The mounting tabs should lock over the ridges on the mounting ring tracks. Okay, now tighten the rings one last time. The disposal should now be self-supporting.

Locking mounting assembly

Attaching the Discharge Tube and P-Trap

Slide the nut (metal flange) over the new discharge tube, insert the rubber gasket into the discharge outlet, and connect it to the disposal with the provided screws. Finger-tighten, and then use the plumber's wrench to tighten it more, being careful not to overtighten.

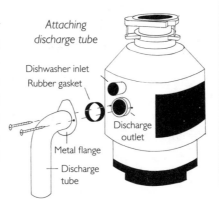

Attaching discharge tube

Dishwasher inlet
Rubber gasket
Discharge outlet
Metal flange
Discharge tube

Replace the p-trap (rotate the disposal to better align with the p-trap, if necessary) and finger-tighten the p-trap nuts, and then tighten with the plumber's wrench, being careful not to overtighten. Place some paper towels underneath the trap.

Attaching the Disposal to the Dishwasher

To connect the disposal to the dishwasher, attach it through the air gap, which is located above the discharge tube. Fasten the clamp to

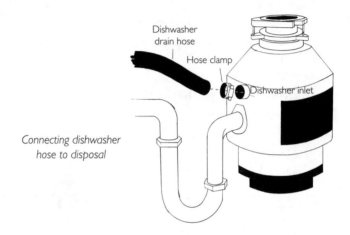

Dishwasher drain hose

Hose clamp

Dishwasher inlet

Connecting dishwasher hose to disposal

secure the drain hose to the dishwasher inlet.

Checking Your Work

Have your helpful friend turn on the power at the main service panel. Turn on the water and the electrical switch at the sink so that the garbage disposal is running. Now check for leaks under the disposal. If there aren't any, then turn everything off.

If There Are Leaks

Don't panic! The problem could be that the p-trap is loose, which is easily fixed by tightening the nuts. And if water is still leaking, then it may be that the disposal is a bit askew, which means that you'll need to remove it and try again.

Tools Needed

New garbage disposal

Manufacturer's instructions

Flashlight

Helpful friend

Towel

Voltage tester

Small, shallow bucket

Plumber's wrench

WD-40

Hand wipes

Washers for p-trap (if necessary) ⭘

Garbage can

Phillips screwdriver

Hammer

Wire stripper (if necessary)

Paper towels

Stand-alone flashlight

Replacing a Kitchen Faucet

on't let your kitchen *sink* to a new low in looks. Get a quick handle on the problem by replacing the old faucet!

There's a dirty little secret to replacing a kitchen faucet that no one (i.e., book authors, online plumbers, and do-it-yourself Web sites) wants to tell you and we know why, because when you find out you'll scream, *"For-get it!"* But we've always been upfront and honest with you, so here goes . . . you have to remove the garbage disposal first.

There, it's out in the open now.

Hold on, sister—if we tell you it can be done, then it can be done. So, let's do this, okay?

Buying a Kitchen Sink Faucet

Replacing a kitchen sink faucet is a job that you want to do only once, and if you buy the right type of faucet, that's what will happen. Style is relative—it's what's inside that really counts (sounds like something Mom would say). The most durable faucet is one that has a disk assembly (a.k.a. its innards) A *disk faucet* costs more, but it typically comes with a very good warranty, so it pays to sink (*hee-hee*) some money into a high-quality faucet.

A *lever handle* is the most popular and practical style of faucet to have in a kitchen because it allows you to be hands-free, if necessary. In fact, you can turn it on and off by just flipping the lever up and down with your elbow, which is a bonus when you've been making meatloaf. And if you suffer with arthritis, a lever is optimal because you don't have to grip it.

When buying a replacement faucet, you need to know the spacing distance of the existing faucet. In other words, how many holes does the sink have for the faucet, and how far apart are the holes? Sinks will have either one hole (a faucet with a lever handle) or three holes (one for the faucet and one for each handle). The holes for the handles are each typically either 4 inches or 8 inches from the faucet. You may think that your sink has one hole, but the escutcheon (a.k.a. base plate) may be covering the other two. So, if you're not sure, stick your head underneath the sink and look to see how many holes exist.

Also, be sure that the new faucet's escutcheon is the right length to cover the holes that you need it to cover. We recommend that you measure the existing escutcheon and bring a tape measure with you to the store to measure the new one. And if the new faucet's escutcheon won't cover all the holes, then purchase matching blank plugs.

Regarding a sink sprayer: if you have one, then you know that a hole exists for it. If you don't have a sink sprayer, look on the sink near the faucet to see if there's a blank plug. If so, you can pop it out and use that hole for the sprayer. And if not, then you can purchase a faucet that has a sprayer in its head.

And one last thing to consider . . . since you'll be removing the p-trap and the garbage disposal, think about buying replacements now, if necessary.

Getting Started

Before doing anything, remove the new faucet and parts from the box, make sure that everything is accounted for, and read the manufacturer's directions. It's important to note that some manufacturers' list of "tools and materials" is for *all* their faucets, so don't panic if some items aren't really needed. *That's why we're here.*

Place a folded towel on the floor in front of the sink so you can kneel on it.

Remove everything from underneath the sink and put it away from where you'll be working, because you'll need as much space as possible for this project.

Lay a towel inside the cabinet to catch any drips and place a stand-alone flashlight near the sink to provide extra light—even if you're doing this during the day.

*F*aucets are either **top mount** or **bottom mount. That sounds very simple, but here's where it can be confusing: For *top-mount* faucets, the nuts that secure the faucet to the sink are located *underneath* the sink. And for *bottom-mount*, the *escutcheons and handles* must be removed to reach the nuts that secure the faucet to the sink. Since most faucets are top-mount, that's what we've installed.**

Shutting Off the Water

Reach behind the garbage disposal and p-trap and turn off the water at the two shutoff valves (hot and cold). If the shutoffs are not there, then turn off the water at the main water valve.

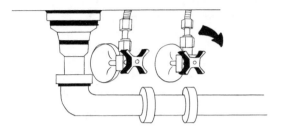

Turning off water at shutoff valves

Turn on the faucet to release any remaining water or pressure that may be in the water lines.

Removing the P-Trap and Garbage Disposal

Place the small, shallow bucket under the p-trap. If the p-trap is white, then it's made of plastic, which means you can probably remove the nuts with your fingers. If the p-trap is metal, then place the plumber's wrench on the nut and turn it counterclockwise. If it doesn't budge,

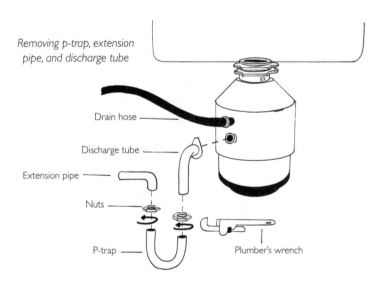

Removing p-trap, extension pipe, and discharge tube

Drain hose

Discharge tube

Extension pipe

Nuts

P-trap

Plumber's wrench

spray WD-40 on all the nuts and wait about 15 seconds before trying again. Once the nuts are loosened, remove the p-trap and dump the contents into the bucket. Place the p-trap on the towel. Be warned that it may smell, as in stink.

Remove the garbage disposal (see page 4).

Removing the Water Lines

Note: If your new faucet came with water supply lines preattached, you can skip ahead with the instructions.

Use the basin wrench to loosen the coupling nuts at the top of the supply lines directly under the sink. You may find that it works best to lie on your back to get a better angle. If so, place a bath mat inside the cabinet for your back to rest on, and remember that if you're lying on your back then you'll be working "backward," so left will be to tighten and right will be to loosen. If the coupling nuts aren't budging, spray WD-40 on them, wait a few minutes, and then try again.

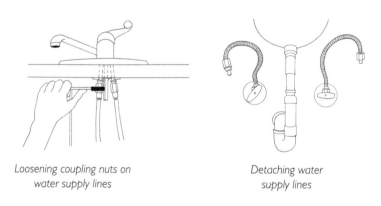

Loosening coupling nuts on
water supply lines

Detaching water
supply lines

You should always use flexible steel water supply lines when installing a new faucet.

Removing the Sink Sprayer

Since you'll be replacing the sink sprayer with the faucet, you can save yourself a little time and trouble by just cutting the sprayer hose with scissors.

Hand-loosen and remove the nut that holds the sprayer in place—it's located underneath the sink, directly below the sprayer. Pull the sprayer up and out, and discard.

Removing the Old Faucet

Use the basin wrench to loosen the mounting nuts that secure the faucet to the sink, which are located directly underneath the sink. Remove the nuts and the washers and discard, along with the mounting nuts.

Now pull the faucet up and out, and dispose. You'll see old plumber's putty, gook, an old gasket, or all of the above. Clean the area completely before installing the new faucet.

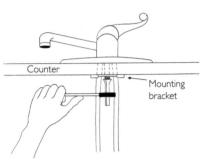

Counter

Mounting bracket

Loosening mounting nut

Note: Some manufacturers require using Teflon tape on the threaded shank of a faucet, but others don't, so, as always, refer to the manufacturer's directions.

Installing the Gasket or Applying Putty

Every faucet needs a gasket or plumber's putty (it's always one or the other) to act as a barrier to water entering the cabinet.

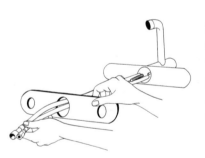

Installing a faucet gasket

If the faucet came with a gasket, slide it over the faucet supply hoses and hose outlet. If the faucet did not come with a gasket, then you'll need to use plumber's putty. The best way to do this is to take some putty in your hand and roll it until it's long, and then apply it to the inside perimeter of the base of the faucet.

Applying Teflon Tape to the Sink Sprayer

Before installing the faucet, you need to apply Teflon tape to the threaded part of the spray inlet connector. It's small, so you won't need much. Wrap the tape counterclockwise (looking at it straight on) until you have two or three layers. Be sure to stretch the tape to get it into the grooves.

Adhering Teflon tape to spray inlet connector

Installing the Faucet

Position the faucet so that the tailpiece and water supply tubing (hot and cold) go into the center hole on the sink. Push down on the faucet. If you used plumber's putty, don't panic when you see the putty oozing out the sides. Just wipe it away with a damp rag.

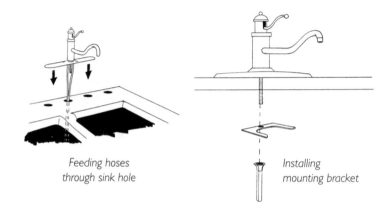

Feeding hoses through sink hole

Installing mounting bracket

Move the faucet so that the spray inlet connector is located at the back and the faucet is positioned correctly. Insert the new washers and mounting nut onto the tailpiece, following the manufacturer's directions. Hand-tighten and then tighten using the basin wrench. Be careful not to overtighten because you don't want to crack the sink.

Reattaching the Water Supply Lines

Wrap Teflon tape (2 or 3 layers) on the hot and cold shutoff values.

Connect the water supply lines to the shutoff valves, hand-tighten, and then slightly tighten with the adjustable wrench.

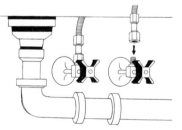

Reattaching water supply lines

Installing the Sink Sprayer

Unravel the spray hose. This will make it easier to maneuver under the sink.

Installing a sink sprayer

Insert the gasket onto the base of the spray holder and then place it over the hole for the sprayer. Insert the hose through the spray holder and through the sink, and pull it gently from below.

Feel for the spray inlet connector on the tailpiece of the faucet. Attach the coupling nut on the hose to the connector and tighten.

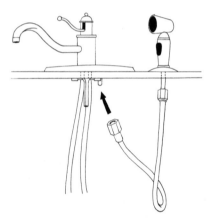

Attaching coupling nut to connector

Replacing the P-Trap and Garbage Disposal

Install the p-trap and garbage disposal following the steps in the reverse order.

Turning on the Water

Remove the aerator on the faucet and sink sprayer by twisting them off in a counterclockwise direction. By doing this before turning the water on, you will keep any debris in the water lines from getting into the aerators.

If the faucet has a ceramic disk, you'll need to turn the faucet to the center on position, *prior* to turning the water supply on because a sudden surge of air may

Removing aerator

crack the disk. Otherwise, keep the faucet in the off position while you turn the water supply on.

Now turn the handle(s) to the center on position. Check above and below for leaks. Replace the towel with a few paper towels and leave them there for a day or two so you can continue to check for leaks. If there is a leak, tighten the mounting nuts.

Replace the aerators.

Tools Needed

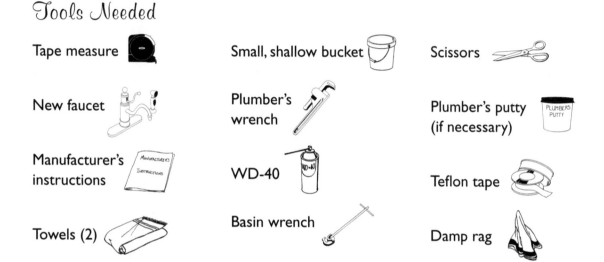

Tape measure

New faucet

Manufacturer's instructions

Towels (2)

Stand-alone flashlight

Small, shallow bucket

Plumber's wrench

WD-40

Basin wrench

Bath mat

Adjustable wrench

Scissors

Plumber's putty (if necessary)

Teflon tape

Damp rag

Paper towels

Replacing a Bathroom Sink Faucet

When Patti downsized from a McMansion to a bungalow, she thought the projects would diminish in size, too. Unfortunately, it seemed that her petite abode's problems were all-the-more noticeable, like the decrepit-looking bathroom faucets. The first project on her list was to replace the one in the guest bath. "My house may be small," Patti said, "but the welcome will be big!"

A faucet that leaks 60 drops per minute wastes 5 to10 gallons of water a day.

Replacing a faucet can perk up any bathroom, no matter the size. And if you're putting your house on the market, then it's a must-do project to avoid losing money down the drain.

Buying a New Faucet

When it comes to faucets, style is relative—it's what's inside that really counts. The most durable faucet is one that has a disk assembly. A *disk faucet* costs more, but it typically comes with a very good warranty.

When buying a replacement faucet, you need to know the spacing distance of the existing faucet. In other words, how many holes does the sink have for the faucet, and how far apart are the holes? Sinks will either have one hole (a faucet with a lever handle) or three holes (one for the faucet and one for each handle). The holes for the handles are typically either 4 inches or 8 inches from the faucet. You may think that your sink has one hole, but the escutcheon (base plate) may be covering the other two. So, if you're not sure, stick your head underneath the sink and look to see how many holes exist.

Also, be sure that the new faucet's escutcheon is the right length to cover the holes that you need it to cover. We recommend that you measure the existing escutcheon and bring a tape measure with you to the store to measure the new one.

Getting Started

Before doing anything, remove the new faucet and parts from the box, make sure that everything is accounted for, and read the manufacturer's directions. It's important to note that some manufacturers' list of "tools and materials" is for *all* their faucets, so don't panic if some items aren't really needed. *That's why we're here!*

Take everything out from underneath the sink and place it inside the bathtub or outside the room, because you'll need all the space you can get for this project.

Place a towel underneath the water supply lines to catch any drips, and place a stand-alone flashlight near the sink to provide extra light—even if you're doing this during the day.

Removing the Old Faucet

Note: If the old faucet is in good working condition, think about donating it instead of throwing it out. Be sure to put all the old parts in a plastic bag as you disassemble it.

Shutting Off the Water

You don't have to turn off the main water valve for this job. Instead, just turn off the water at the hot and cold shutoff valves directly under the sink (if the shutoffs are not there, then turn off the water at the main shutoff valve). Remember, *righty tighty, lefty loosey*. Turn on the faucet to release any remaining water or pressure that may be in the water lines.

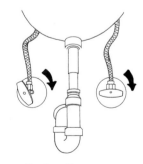

Turning off water at shutoff valves

Removing the Water Supply Lines

Note: If your new faucet came with water supply lines preattached, you can skip ahead to Disconnecting the Pop-up Mechanism.

Use the basin wrench to loosen the coupling nuts at the top of the supply lines directly under the faucet. You may find that it works better to lie on your back to get a better angle. If so, place a bath mat inside the cabinet for your back to rest on, and remember that if you're lying on your back, then you'll be working "backward," so left will be to tighten and right will be to loosen. If the coupling nuts aren't budging, spray WD-40 on them, wait a few minutes, and then try again.

You should always use new flexible steel water supply lines when installing a new faucet.

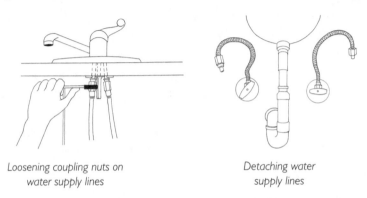

Loosening coupling nuts on water supply lines

Detaching water supply lines

Disconnecting the Pop-up Mechanism

The clevis strap is attached to the faucet's lift rod, so you'll need to remove it before you take out the faucet. This can be difficult to see, so feel behind the p-trap for the clevis strap and clevis screw.

Loosen the clevis screw with your fingers. If it won't budge, spray it with WD-40, wait a few minutes, and then use an adjustable wrench to loosen it. Once it's loose, pull out the screw and put it in your pocket. The faucet's lift rod is now free.

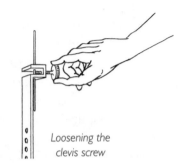

Loosening the clevis screw

Removing the Faucet

Use the flathead screwdriver to loosen and remove the mounting nuts that secure the faucet to the sink (the flathead of the screwdriver fits into the slot of the mounting nuts that secure the faucet to the countertop). These are located directly underneath the sink. Remove the nuts and the washers.

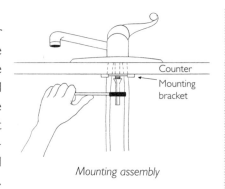

Mounting assembly

Now pull the faucet up and out, and dispose. You'll see old plumber's putty, gook, an old gasket, or all of the above. Clean the area completely before installing the new faucet.

Installing the New Faucet

Installing the Gasket or Applying Putty

Not all faucets are manufactured alike, and we're not just talking about looks. Some faucets come with a gasket (either a rubber O-ring or a plastic template), which acts to prevent water from seeping under the faucet into the cabinet), and others don't. If the new faucet does not have a gasket, then you'll need to use plumber's putty to act as the water barrier. The best way to do this is to take some putty in your hand and roll it until it forms a long piece, and then apply it to underneath the faucet so that it will adhere and seal to the counter.

Installing the Faucet

If the faucet did not come with water supply lines preattached, then connect the lines to the threaded inlets of the faucet.

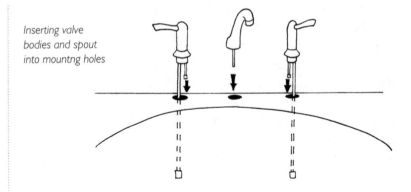

Inserting valve bodies and spout into mountng holes

Position the faucet so that the tailpieces (a.k.a. valve bodies) go into the holes on the sink, making sure that the cold water handle is on the right of the faucet and the hot water handle is on the left. Push down on the faucet. If you used plumber's putty, don't panic when you see the putty oozing out the sides—just wipe it away with a damp rag.

Securing the Faucet

If there are handles, position them as you want them to rest (i.e., handles facing forward, facing out, etc.). Use the new washers and mounting nuts provided with the faucet. Insert a washer onto each tailpiece (including the threaded shank of the spout), and then insert a mounting nut onto each one. Hand-tighten and then tighten using the basin wrench. Be careful not to overtighten because you don't want to crack the sink.

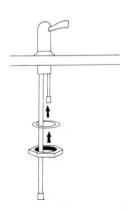

Installing washers and nuts

Reconnecting the Pop-up Mechanism

This may be the only tricky part of the entire project, because it's difficult to reach and it's dark in there. Therefore, you may find it works best to set up a stand-alone flashlight and lie on your back on top of the bath mat to get this done.

Line up the lift rod with the clevis strap and insert the clevis screw. Hand-tighten first and then tighten with an adjustable wrench.

Attaching the Water Supply Lines to the Water Supply Valves

Lining up lift rod with clevis strap

Lift rod

Clevis strap

Wrap Teflon tape (2 or 3 layers) on the hot and cold shutoff valves and the threaded shank of the faucet.

Connect the water supply lines and hand-tighten the coupling nuts. Use the adjustable wrench to tighten the nuts a bit more.

Note: Some manufacturers have the hot and cold supply lines connected inside the faucet, while others require you to connect them in what's called a "T-joint." Follow the manufacturer's instructions for connecting all supply lines.

Some manufacturers require using Teflon tape on the threaded shank of a faucet, but others don't, so, as always, refer to the manufacturer's directions.

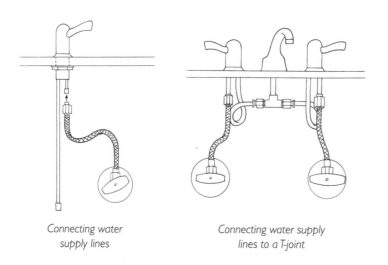

Connecting water supply lines

Connecting water supply lines to a T-joint

Turning on the Water

Remove the aerator on the faucet by twisting it off. Doing this before turning the water on will keep any debris from the water lines getting into the aerator.

Even if the new faucet came with a new drain, you don't have to replace the old one, unless you're changing from stainless steel to copper, etc. If so, follow the manufacturer's instructions for installation.

If the faucet has a ceramic disk, you'll need to turn the faucet to the center on position, *prior* to turning the water supply on, because a sudden surge of air may crack the disk. Otherwise, keep the faucet in the off position while you turn the water supply on.

Check above and below for leaks. Replace the towel with a few paper towels, and leave them there for a day or two so you can continue to check for leaks. If there is a leak, tighten the mounting nuts.

Replace the aerator.

Removing aerator

Tools Needed

Tape measure

New faucet

Manufacturer's instructions

Towel

Basin wrench

WD-40

Bath mat

Flexible steel water supply lines (if necessary)

Adjustable wrench

Flathead screwdriver

Teflon tape (if necessary)

Plumber's putty (if necessary)

Damp rag

Stand-alone flashlight

Paper towels

Replacing a Bathroom Vanity with Countertop Sink

Your master bathroom is where you see your birthday suit every morning ... in all its glory. So, why not surround yourself with a beautiful sink and cabinet? And soft lighting, of course.

If you're going to spend some money to improve your home, there's no room that will give you more bang for your buck than a bathroom.

Now, here's the catch . . . if you're going to replace the vanity, then you'll probably want to replace the faucet. And since you're taking out the vanity, you may want to replace some of the cracked tiles. And, oh, maybe the old toilet should go, too.

We're just giving you advance warning.

Buying a New Vanity

Bathroom vanities can be purchased ready-made or made-to-order. If you're looking to save time and money, then we strongly recommend the ready-made variety.

Before buying a vanity, you'll need to know the height, width, and depth of the old vanity, including the countertop sink, which you'll have to purchase separately.

The common *height* for a new vanity is 33½ inches, which is probably a little bit taller than the old one you have. This is important to note (especially if you have a wall-to-wall mirror) because this measurement does not include the height of the countertop and the 4 inches for the backsplash.

The most common *widths* for vanities are 30 inches, 36 inches, 48 inches, and 60 inches (this size typically includes two sinks). There

Ready-made vanities come in a box and are cash-and-carry. However, you'll probably have a 2- to 4-week wait for the countertop sink to be ready for pickup at the store.

are smaller ones (18 inches and 21 inches), which are designed for powder rooms and typically come with the countertop sink.

No matter the height or width, the standard *depth* for a vanity is 22 inches.

One more important feature that we think is a deal-breaker . . . the vanity must be backless. If the vanity doesn't have this feature, you're in for a lot of extra measuring and sawing.

Buying a New Countertop Sink

If you're using the old countertop sink, then you can skip this section. If not, be aware that this is where it can get pricey.

You know you'll be paying more if you choose granite or Corian over faux marble, but what you may not know are all the additional charges that can accrue. For example, every countertop sink comes with a backsplash, but if you want side splashes, you'll be paying extra. There are dripless edges that carry no charge, but every other type of edge comes with an additional cost. And then there are the holes in the sink—if you need one hole for the faucet instead of the typical three holes, you'll be paying more. And the list goes on.

Edges and colors are a matter of choice, but side splashes are a matter of practicality. A backsplash is designed to protect the adjacent wall from water "splashing" onto it. Therefore, you'll need to order one or two side splashes, depending on whether the vanity rests against one or two side walls.

You need to know the measurement of where the existing bowl is located, because some are off-center, depending on the location of the plumbing. Therefore, measure starting from the left of the bowl to the center of the drain hole.

Getting Started

There are a couple of things you need to know prior to getting started on this project. One is that you will more than likely do some wall damage when removing the cabinet and countertop, so schedule time for spackling and painting.

The other thing is, don't be surprised if after removing the cabinet you find that the flooring is different from what's in the rest of the bathroom. In fact, you may notice that there's no tile or linoleum, just subflooring. The builder or previous homeowner tried to save a little money by not removing the vanity when installing the flooring. If this is the case, then you'll need to install flooring that will be the same height as the original so that the new cabinet will be level. And if you're going with a smaller cabinet, then you'll need to install matching flooring.

You'll need to remove the hot and cold water supply lines and the p-trap (see "Replacing a Bathroom Sink Faucet"). The faucet can stay attached because you can disconnect it after removing the countertop, if necessary.

Depending on the size of the old and new vanities, you may need to remove the bathroom door to get the old one out and the new one in.

And of course be sure to remove everything from underneath the sink and from inside any drawers, and properly dispose of any old medicines.

Removing the Old Countertop Sink

If you can reuse the old countertop sink, then great—one less thing that goes to the dump! If not, then here are some easy steps to follow.

The first thing you need to do is to remove the caulking between the wall and the countertop. Use a sharp utility knife to cut through the caulk. Then insert a metal putty knife between the wall and the countertop and tap it with the mallet to create a wedge.

You may be surprised to learn that the countertop sink is attached to the vanity with only some adhesive . . . no fasteners. If you'll be reusing the countertop, be very careful when doing the next steps to avoid breaking it.

The best way to loosen the countertop sink is to use a mallet and hit it from below, inside the vanity. Knock it a few times until you see movement. Try lifting up the countertop sink. If it's not budging, you'll need to knock it with the mallet a few more times.

Separating countertop from wall

Once it's loosened, remove it and set it in a safe, temporary spot, such as the bathtub. Depending on the size and make of the countertop sink, you may need a helpful friend to aid you in lifting it off.

Removing the Old Vanity

The best way to remove the old vanity is to lighten the load first. This means that you should remove all the drawers and doors. To remove the drawers, simply pull them out and lift them off their tracks. To remove the doors, open them to expose the hinges and use the Phillips screwdriver to loosen all the screws.

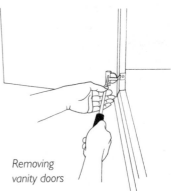

Removing vanity doors

Now that the inside of the vanity is completely exposed (*woo-hoo*), you'll see that the only things keeping it attached to the wall are a few screws.

Use the Phillips screwdriver to remove all the screws and put them in a plastic baggie in case the new vanity is missing any (they should be the same size).

Once all the screws, doors, and drawers have been removed, lift out the vanity and dispose of it properly.

Patch and paint any damaged areas.

Installing a New Vanity

As noted above, the bathroom door more than likely needs to be off its hinges in order for you to get the new vanity inside.

Place the vanity into the correct spot and place a level on top. If the vanity isn't level, first check to see if it's resting on a piece of debris. If not, adjust the built-in levelers in each of the legs, if provided, or place a shim underneath to make it level. Use the utility knife to cut away any part of the shim that's exposed.

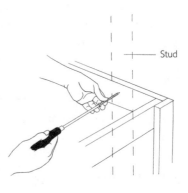

Stud

Securing vanity to wall studs

To secure the vanity to the wall studs, use the Phillips screwdriver to insert all the screws.

Attaching Vanity Doors and Inserting Drawers

Before installing the doors and drawers, attach all pulls and knobs first, using a Phillips screwdriver.

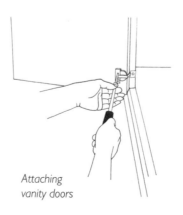

Some vanity doors come with preinstalled hinges that easily snap into place, while others require the use of a Phillips screwdriver. Follow the manufacturer's directions for installation.

Attaching vanity doors

Installing a New Countertop Sink

Note: You can install the faucet prior to installing the countertop. Just be sure to keep the water supply lines out of the way before laying it on top of the vanity.

Place a bead of silicone around the top of the vanity where the countertop will sit. Depending on the

Applying silicone to vanity top

Installing the countertop sink

size of the countertop sink, you may need a helpful friend to place it on top of the vanity.

If you have any side splashes and they're not attached to the countertop, apply silicone to the backs and press in place. Apply some pieces of painter's tape across the side splashes to keep them in place. Remove the tape within 2 hours.

Using the silicone recommended by the countertop manufacturer, apply a bead around the perimeter, where the countertop meets the wall. Run your finger around it to make a smooth line, if necessary, and wipe away any excess.

Allow 24 hours for drying before using the countertop sink.

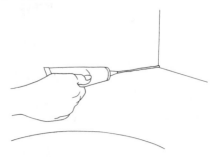

Applying caulk where the countertop meets the wall

Tools Needed

Tape measure

New bathroom vanity with countertop sink

Metal putty knife

Manufacturer's instructions

Rubber mallet

Helpful friend (if necessary)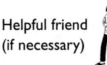

Phillips screwdriver

Level

Shims (if necessary)

Utility knife

Silicone caulk and caulking gun

Painter's tape

Replacing a Toilet

This is one of those projects that will earn you bragging rights at the water cooler or car pool and not because it's hard to do . . . it just sounds like it. But don't worry—we'll never tell!

Toilets are basically indestructible; in fact, toilets can typically last up to fifty years. So, why would you ever need to replace one? There are several reasons: (1) if the porcelain cracks and water is leaking out; (2) if the color of the toilet doesn't match the rest of the updated bathroom; and (3) if it's an old toilet, because it uses too much water per flush.

The EPA recommends that all older toilets be replaced with water-saving models, also known as low-flush toilets. The difference in water saved would be astronomical—up to 2 billion gallons a year.

And don't say that a new toilet just won't flush as well as your old one because that logic doesn't hold water anymore (*hee-hee*).

Okay, so you agree that you need to save on your water bill by buying a new toilet, but why shouldn't you just hire a plumber to do the job? Because you *can* do this project, and you *will* save hundreds of dollars in the process.

Buying a New Toilet

Choosing a Toilet

The first decision to make about a toilet is the shape. Toilets come in round and elongated. A round toilet takes up less space, so it works well in a powder room.

Another feature to consider is height. You'll probably never replace the toilet you're installing, so with that in mind, consider your future needs. For example, you know you're not going to grow taller, but the "comfort height" toilet (a.k.a. "chair height") is also wonderful for people with knee, back, or hip problems. Also, toilets come in different lengths, so be sure to measure correctly (see below) for a proper fit.

In addition to the typical two-piece version, toilets now come in one-piece models. What's the big deal? Well, some people think the one-piece model looks sleeker than its predecessor, and because it's one piece, it's easier to install. What we learned is that because it's one piece, it's much heavier, thereby making it more difficult to install . . . even with the help of a friend. The good news is that Kohler has a model, the Persuade toilet, which looks like a one-piece but is actually two pieces, so it has the best of both.

Something that you never had to think about before is the different choices of flushing systems. All new toilets are low-flush, but now they provide such a powerful flush that you'd think it was a toilet from the 1960s. Kohler states that a toilet that flushes 1.28 gallons of water or less, compared to a 3.25-gallon toilet, can save up to 22,000 gallons of water per fixture, per year.

And of course there's color. You'll pay extra for anything other than white and off-white. Who knew there were so many decisions to make when choosing a toilet, right?

Dimensions of Space

Now that you know the height, width, shape, and color you want for your new toilet, you need to measure for the correct size. Actually, you're not going to be measuring the toilet, but rather the space for it.

Use the tape measure to determine the distance from the wall behind the toilet to the center of one of two closet bolts (a.k.a. flange bolts). These are the bolts at the base of the toilet and have plastic caps on them. Now measure the distance from the closet bolts to the wall or vanity on each side. These measurements will ensure that you'll have enough space to install the toilet—definitely something you want to know ahead of time.

Measuring space for new toilet

Purchasing a New Flange and Wax Seals

When you purchase your new toilet, be sure to buy a new flange so that you can replace the old one. You don't want to put a new toilet on a twenty-year-old flange because flanges don't last forever. Also, be sure to purchase two wax seals (one with bolts), because if you make a mistake when installing the toilet, you'll have to replace the wax seal. Better to return the unused one later than to put the toilet project on hold while you run to the store!

Getting Started

First, we think it's a good idea to clean the toilet and surrounding area before removing it, because you are going to be up close and personal with it.

Second, we recommend that you place old towels around the toilet, because this job can get messy.

Third, remove the new toilet from the box without damaging the box and place the box outside, with the opening at the top, for trash pickup. You'll be putting the old toilet inside it either to take it to the dump or leave it for trash pickup. Speaking of which, make sure that your waste disposal company will take an old toilet *before* putting it outside.

And last but not least, clear a path to where you will be taking the old toilet. It's heavy, and you don't want to be tripping over things on your fast exit from the house.

Now you're ready!

Removing the Old Toilet

Removing the Water

Turn off the water at the toilet shutoff valve and flush the toilet—the purpose is to remove as much of the water in the tank and bowl as possible. Take off the tank lid and place it in a safe location. Wearing the rubber gloves, remove any water that's left inside the tank with the sponge. Use the sponge

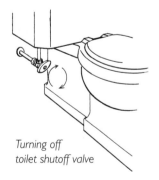

Turning off toilet shutoff valve

Check the flooring around the toilet. If it's squishy to the touch of your feet, you may need to have the subfloor (the unfinished wood that lies underneath tile, linoleum, wood flooring, etc.) replaced. Contact a reputable contractor immediately. And don't pout. Be grateful that the toilet hasn't fallen through the floor . . . yet.

*I*t's best to wear rubber gloves while removing the water in the tank, not because the water is dirty, because it's not. It's because the flapper can leave a slimy residue on your hands (and clothing) that is difficult to remove.

to remove any water that's left inside the toilet bowl and throw out the sponge.

Disconnecting the Water Supply Line

Place the small, shallow bucket under the water supply line. If you will *not* be replacing the supply line, then use your fingers to loosen the coupling nut at the base of the tank and leave it attached at the shutoff valve. If you *will be* replacing the supply line, use an adjustable wrench to loosen the nut at the shutoff valve. If the nut won't budge, spray it with WD-40, wait a few minutes, and then try again. Once the supply line has been disconnected (completely or partially) there will be a little bit of water that comes out.

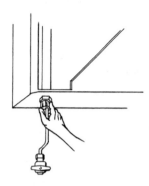

Loosening the coupling nut

Note: If your water supply line is made of rubber, replace it with a flexible steel model to avoid future cracks and leaks.

Remove the fill valve (inside the tank) and discard. This will allow the remaining water, if any, to flow into the bucket. All the other innards can stay inside the tank.

Removing the fill valve

Removing the Tank

Place the adjustable wrench on one of the nuts at the exterior base of the tank. With the other hand, use the flathead screwdriver to loosen the bolt inside the tank, while making sure that the nut on the other end doesn't move. Remove the bolt, washer, and nut and repeat with the other set. Lift off the tank and place it near the lid.

Removing the Flange (Closet) Bolts and Nuts

Pry off the caps to the flange (closet) bolts with the flathead screwdriver. Use the adjustable wrench to remove the nuts on the bolts. If they won't budge, spray them with WD-40, wait a few minutes, and then try again. If the nuts still won't move, use the hacksaw to cut off the bolts as close to the nuts as possible. Discard the nuts and covers.

Prying off flange (closet) bolt caps

Removing the Toilet Bowl

Now is the best time to open any doors for your quick exit. With the help of a friend, lift the bowl off the floor (you may need to rock it back and forth) and put it down nearby. Unscrew the flange and discard.

Immediately stuff the rag into the sewer hole while holding the string. The rag will prevent sewer gases from entering the bathroom, and the string will prevent the rag from entering the sewer system.

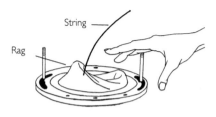
Unscrewing the flange

Now pick up the toilet and place it inside the box that's outside.

Clean the floor by using the metal putty knife to remove any residue so that the new toilet will have a tight fit.

Installing the New Toilet

Installing a New Flange and Flange Bolts

Place the new flange over the sewer hole. If the rag is in the way, take it out, put the flange over the hole, and then stuff the rag back inside.

Insert each flange bolt into the holes in the flange, and move them so that they're seated at the 3 o'clock and 9 o'clock positions. Pinch off a little bit of the wax seal (about the size of a quarter) and put it on the heads of the bolts (the threaded ends of the

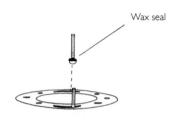
Inserting flange (closet) bolts into flange holes

*A*ll new toilets come with the internal parts of the tank already installed.

bolts are sticking up) so that they won't move when you're placing the toilet on top.

Installing the Wax Seal

Lay the toilet bowl on its side on the bath mat. Remove the wax seal from its box and remove its plastic case, if applicable. Take the wax seal and place it onto the flange so that the tapered end is facing you. Press down on it.

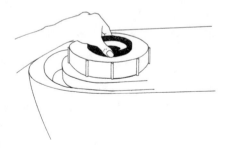

Installing the wax seal

Installing the Bowl

Remove the rag from the sewer hole.

With the help of a friend, lift the bowl and carefully place it onto the flange so that the holes are properly positioned over the closet bolts.

Note: If you were unable to successfully adhere the bowl to the flange, you will need to remove the toilet, take off the wax seal, and put a new wax seal on. Once a wax seal has been compromised, it cannot be used.

Before you secure it, place a level on top of the bowl to see if you need to add a plastic shim underneath.

Place a metal washer over each closet bolt and then a nut over each. Hand-tighten one nut and then the other. Then use an adjustable wrench to tighten each, being *very* careful not to overtighten because you could easily crack the porcelain. Use a hacksaw to cut off the tops of the bolts, if necessary. Attach the caps.

You're almost done!!!!

Attaching the Tank

Lay the tank on the bath mat. Push the gasket onto the threaded extension on the exterior base of the tank (the tapered end of the gasket will fit into the inlet opening in the bowl).

Installing gasket on tank base

Place the tank onto the bowl, being careful to position the holes in the tank over the holes in the back of the bowl, and the gasket into the inlet opening. Once that's done, you can secure the tank.

Now, pay attention here, because it's vital that you put the rubber washers in the right places.

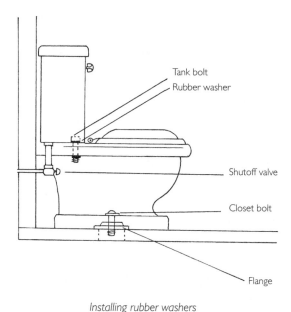

Tank bolt
Rubber washer
Shutoff valve
Closet bolt
Flange

Installing rubber washers

Put a *rubber* washer on each of the bolts and insert the bolts into the holes on the interior base of the tank. The rubber washers and the heads of the bolts will be inside the tank. The threaded parts of the bolts will be on the exterior side of the tank.

Now, on the exterior side of the tank, put a *metal* washer on the bolts first, and then the nuts. Hand-tighten the nuts.

To secure the tank to the bowl, place the adjustable wrench on one of the nuts at the exterior base of the tank. With the other hand, use the flathead screwdriver to tighten the bolt inside the tank, while making sure that the nut on the other end doesn't move. Repeat with the other bolt and nut. *Be very careful not to overtighten or you may crack the porcelain!*

Manufacturers may differ with parts and terms; therefore, read the product's instructions.

Attaching the Water Supply Line

Wrap Teflon tape (2 or 3 layers) on the threaded end of the water supply valve and on the threaded end of the tank's water inlet.

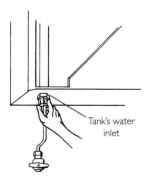

Tank's water inlet

Reconnecting the water supply line

Reconnect the water supply line at the exterior base of the tank and at the water shutoff valve. Hand-tighten the coupling nuts first, and then use an adjustable wrench to tighten them more. *Be careful not to overtighten at the base of the tank to avoid cracking the porcelain.*

Place the bucket underneath the water supply line and flush the toilet. If water is leaking, tighten the coupling nuts.

Installing the Toilet Seat

Position the seat so that the holes on the hinges align with the holes on the back of the bowl. Put a screw into a hole and thread a nut from below onto it. Hand-tighten the nut. Repeat with the other screw and nut.

Use a flathead screwdriver to hold the screw in place while you use an adjustable wrench to tighten the nut from below. Repeat on the other set. Snap the caps down.

Applying a Sealant

It's best to wait a day or two for the toilet to settle before applying a sealant around its base.

Cut off the tip of the caulk tube at about a 45-degree angle and about ¼ inch from the top. Apply a thin coat of the caulk around the base of the toilet, making sure to leave some space around where the closet bolts are located. This space will allow any water that leaks from the toilet to dissipate across the floor instead of leaking below the toilet into the subfloor. Wipe away any excess caulk.

Tools Needed

Tape measure

New toilet

New flange

2 wax seals (one with bolts)

Cleaning products

Towels

Rubber gloves

Large sponge

Teflon tape

Manufacturer's instructions

Small, shallow bucket

Adjustable wrench

WD-40 (if necessary)

Flexible steel water supply line (if necessary)

Large flathead screwdriver

Hacksaw (if necessary)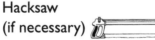

Helpful friend

Rag (with string tied at one end)

Metal putty knife

Bath mat

Level

Plastic shims (if necessary— found in plumbing stores)

Caulk (not heavy-duty adhesive) and caulking gun

Electrical

We know that working with electricity can be intimidating, but we're hoping that the ugly chandelier hanging in the dining room, or the ceiling fan that's been broken for three years, or the need for new outdoor lighting will be enough of a catalyst to get you to turn the page to see just how easy *and* interesting electrical projects can be.

And if you're worried about your safety . . . good. We certainly are, and that's why we provide safety tips and detailed instructions for each project.

So, *lighten* up. You can do it!

Replacing an Exterior Light Fixture

*K*nock, knock. Who's there? Well, maybe you'd know if you replace the outdoor lights!

If you want to give your aging home a facelift, don't just stop at painting. Replace the old rusted exterior light fixtures with new models—that will take years off your house.

Install a timer for the porch lights so you'll always come home to a lighted entrance, which will provide you with better safety. And you'll save electricity, too.

Buying Exterior Light Fixtures

Before you purchase new outdoor light fixtures, you need to know how many need to be replaced (don't forget the one for over the garage door) and the proper size (too small or too large will look too odd). Also, when considering size, keep in mind that the new light fixture should not be larger than the mounting facade (the cover for the mounting hole), which is made of wood, vinyl, or aluminum.

Use the tape measure to get the height and width of the existing exterior lights and the mounting facade and bring the information with you to the store so that you can get the correct replacements.

And while you're shopping for light fixtures, pick up some CFL (compact fluorescent) bulbs of the proper wattage—the information will be located on the new light fixture, as well as on its exterior packaging.

Getting Started

No matter what type of light fixture you buy, the main components for this project should be the same: (1) the exterior junction box, which is attached to the house and contains your home's electrical wires; (2) two anchor screws, which come out of the junction box;

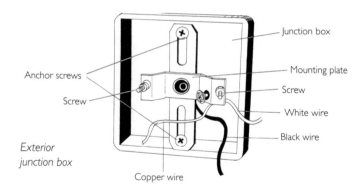

Junction box

Anchor screws

Screw

Mounting plate

Screw

White wire

Black wire

*Exterior
junction box*

Copper wire

(3) a mounting plate that fits over the two screws and onto the junction box; (4) another pair of screws that are inserted into the mounting plate and which attach to the light fixture; and (5) the light fixture.

Open the box and check that everything has been included. Read the manufacturer's instructions.

Turning Off the Electricity

Flip on the switch to the exterior lights. Turn off the power to the light fixture at the main service panel. Have your helpful friend tell you when the light is off.

Note: If your home's circuit breaker box is not properly labeled, be sure to turn off all computers and televisions in the house before flipping any breakers or removing any fuses.

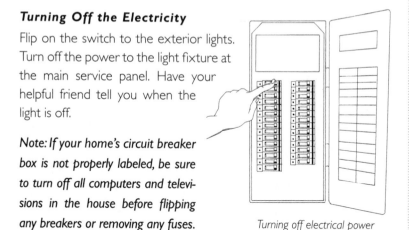

Turning off electrical power

Removing the Old Light Fixture

If necessary, set up a stepladder or ladder near the light fixture, using the appropriate safety measures.

Loosen the set screws that secure the glass housing. Remove the glass housing and the lightbulb and put them in a safe place.

Support the old light fixture while loosening the screws or decorative nuts (a.k.a. acorn nuts), and put the screws in a secure place (e.g., your pocket). Wipe away any dust or cobwebs, if necessary.

If the old light fixtures work, don't throw them out. Instead, recycle by donating them to a thrift store or a Habitat for Humanity ReStore in your area.

Detaching the Wiring

Once the light fixture has been moved away from the exterior wall, you'll be able to see the wiring.

Now you need to double-check that the electricity was turned off. Touch a wire with the voltage tester. If the tester emits a constant beeping noise with a flashing light, then it means that the power was not turned off. Keep flipping the circuit breakers until the electricity has been turned off. Be sure to create a circuit map later.

Take a look at how the wires are connected—they should be black to black, white to white, and green or bare copper to green or bare copper. But that's not always the case. The rule of thumb is that since the wiring works, you'll have to reattach the wires from the new light fixture to the house wires the exact same way. So, draw on a piece of paper exactly how the wires are connected.

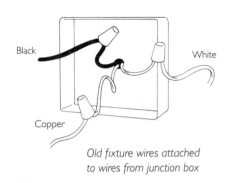

Old fixture wires attached to wires from junction box

Note: If you don't see a green or bare copper house wire, then you need to stop. All new light fixtures are manufactured with a grounding wire, which needs to connect to the home's grounding wire. So, button up the light fixture and hire an electrician to have a grounding wire installed to that box, as well as to check your home's entire electrical system.

Twist off all 3 of the wire nuts and loosen the wires. The old light fixture is now completely disconnected. Hand it to your helpful friend. Almost every new light fixture will fit onto the old mounting plate. Save the new mounting plate and hardware for possible future use.

If you're nervous about stripping a wire, buy a piece of wire from the hardware store to practice on.

Stripping the Wires

To make a good connection, you'll need about ½ inch of exposed wire. If any of the wires, either from the exterior junction box or the light fixture, need more bare wire exposed, use the wire stripper to remove some of the insulation. Simply place the end of the wire into the proper hole of the wire stripper and the proper length that you want stripped. Clamp down and pull off the insulation.

Connecting the Wires

It's always best to connect the grounding wires first. Take the grounding wire from the light fixture and the grounding wire from the exterior junction box and join by twisting a wire nut on top. Finger-tighten.

Take the white wire from the light fixture and the white wire from the exterior junction box, and join by twisting a wire nut on top. Finger-tighten. Repeat the same procedure with the black wires.

Push the wires into the exterior junction box and fit the mounting facade over it.

Note: You have to install this prior to installing the light fixture.

Insert the screws into the new mounting plate and place the plate over the junction box, aligning it so that the holes fit over the screws. Twist the nuts onto the screws and tighten with an adjustable wrench.

Secure the new light fixture with the acorn nuts. Install a CFL bulb and attach the glass housing with the set screws.

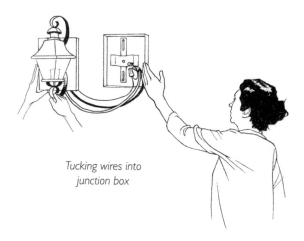

Tucking wires into
junction box

Tools Needed

Tape measure

Stepladder
(if necessary)

New exterior
light fixture

Manufacturer's
instructions

CFL bulb

Voltage tester

Paper
and pencil

Helpful friend

Phillips
screwdriver

Wire stripper

Adjustable
wrench

Installing Landscape Lighting

Why let the sun steal the spotlight from your home every evening? Install landscape lighting to show it off both day *and* night.

Installing landscape lighting, whether it's solar powered or low voltage, is an easy way to highlight your yard's beauty at night. But for those of you whose yards aren't quite ready for primetime, think instead of landscape lighting as a safe way to get you and your guests to the front door.

There are two types of landscape lighting: (1) solar powered, and (2) low voltage. *Solar-powered* lighting is a cinch to install and receives its energy from the sun (with a little assistance from a battery), so it's free to operate, except for the cost of replacement batteries. There are some drawbacks, one of which is that solar-powered lighting requires a sunny location, so if the area you want to light up at night is shady during the day, then solar-powered lighting will not work for you. Another somewhat negative attribute is that the solar-powered models can cast a bluish light.

The benefit to using *low-voltage* landscape lighting is that the area can be sunny or shady. It can create a more dramatic look for your house than solar-powered; you can put this type of lighting on a timer, which can save energy; and it's not difficult to install—really! The downsides are that it costs money to operate and its location is often decided by the distance from the transformer (a.k.a. power source box).

Solar-Powered Lights

Buying Solar-Powered Lights

When purchasing landscape lighting, you definitely get what you pay for. The *lower-grade* product consists of plastic top housing and plastic lens and pole; the *better grade* has plastic or metal top housing with plastic lens and pole; the *best grade* has metal top housing with glass lens and metal pole; and the *premium grade* has cast metal top housing with glass lens and metal pole. We strongly recommend that you don't purchase all plastic pieces. We discovered upon opening a box of six solar lights with clear plastic lenses that one was already damaged, one broke while assembling it, and another broke when a neighbor's dog ran into it. That left three. We think it's worth spending the extra money to buy the best or premium grade product.

Landscape lighting comes in packages of six, eight, or twelve. But you don't always need all the lights. In fact, we've seen too many homes that look like airport runways because people used too many lights to illuminate their yard. Jiawei, a manufacturer of solar lighting, recommends that you use 1 to 2 lights for every 5 to 6 feet.

Party lights are a different story.

Getting Started

Open the box and inspect the lights. Next, take an accounting of all the other parts. If any are damaged or missing, place everything back in the box and return it.

If all the parts are good, follow the manufacturer's instructions for assembly. One thing that is not mentioned in the instructions is that you should be careful to avoid overtightening any plastic parts, because, as mentioned above, they crack *very easily*. It's also important to note that some instructions will be written and some will just be illustrations, so if you're experiencing difficulty assembling the lights, contact the manufacturer via its 800 number.

Assembling Solar-Powered Lights

Note: Manufacturers and models may vary slightly in assembly.

Attach the glass diffuser to the top housing by twisting it into place. Look on the bottom of the top housing for the on/off switch and flip it to the on position.

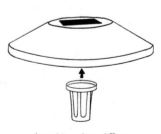

Attaching glass diffuser

Attaching glass lens

Attaching light to mounting pole with ground spike

Now attach the glass lens to the top housing by twisting it into place.

Push the ground spike into the mounting pole. Attach the light to the mounting pole.

Installing Solar-Powered Lights

It's best to do this project first thing in the morning so that the sun can power the photocell. It typically takes 10 to 12 hours for the cell to charge, with the switch in the on position.

Lay the assembled lights where you think they should be located and then take a step back to give the area a once-over. Move the lights, if necessary.

Push the stakes into the soil, making sure that the mounting pole, not the spike, is exposed.

Check your handiwork after sunset to see if you will need more lighting and if the location of the lighting is effective.

Installing light into ground

Tools Needed

Solar-powered landscape lighting

Manufacturer's instructions

Low-Voltage Lights

We totally understand your nervousness about installing low-voltage lighting, because we felt similar butterflies. That was until we did some research and actually installed the lights. Today's low-voltage lights are much safer and so much easier to install than older models that we promise they will cast a new light for you on doing this project.

Buying Low-Voltage Lights

Before buying lighting, you need to know what you want it to do for your home—spotlight a tree or provide light for a walkway—and where you're going to place it—under a tree, in front of bushes, along a walkway.

The best way to determine this is at night. Grab some flashlights (borrow some from neighbors), turn them on, and stand them up in the sections you think you want lit. Make a note of the areas that need lighting and whether you need just path lights, spotlights, or both. You may want to draw the area to scale on graph paper to get a better visual.

As with most other products, when buying low-voltage lighting, you get what you pay for. We bought low-end and high-end lights, and what we found was that the more you pay, the better the product, the less assembling is required, and the better the instructions. In fact, the less expensive product came with just a pictorial, no words. It left *us* speechless!

No matter how much you decide to spend on the lights, make sure that the product has the Underwriters Laboratories (UL) approval.

And just as with solar-powered lights, don't feel obliged to use every light in the box. You don't want your home to be lit up like an airport runway.

Getting Started

This project will require you to do some very shallow digging, but even minor digging can be dangerous, especially if you don't know where the utility lines are buried and how deep. So, in order to prevent hitting a utility line, you need to call 811 about a week before installing the lights;

811 is the new national phone number for Call Before You Dig. The operator will route your call to your local utility companies, which will then send out a professional locator to put flags or temporary paint lines on your property that demarcate the different underground utilities. If you don't have this done *and* you hit a line, you will be charged by the utility company for the repair, and it won't be pretty.

Because the low-voltage transformer is being used outdoors and therefore may be exposed to rain, sleet, or snow, it's important that it be plugged into a ground fault circuit interrupter (GFCI) to prevent electrical shock. So, replace the outdoor receptacle if necessary (see page 74).

Open the box of lights, check for all the parts, and read the manufacturer's instructions.

Assembling Low-Voltage Lights

Follow the manufacturer's directions for assembling the lights, which will probably be very similar to the directions for assembling solar-powered lights listed above. As we stated, if you bought inexpensive lights, you'll be spending a good amount of time assembling them.

Laying the Lights and Cable

Referring to the diagram you drew, do a soft (semipermanent) installation of the lights by inserting the stakes into the ground. Then roll out the cable and lay it near the lights. Because you're going to be placing the cable underground (about 6 inches), be sure not to place it near a utility line.

Connecting the Lights with the Cable

Older models of low-voltage lighting required you to splice all the cable wires. Now manufacturers have created a way to do this without you even realizing what's being done. Here's how.

When buying a GFCI for outdoor use, be sure to get one with an "in-use cover" (e.g., a plastic or metal shield).

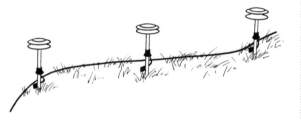

Semipermanent installation of lights

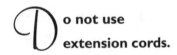
Starting at the end of the cable farthest from where the transformer will be installed, remove the top of the connector. Place the cable that's in front of the light inside the wire channel in the connector.

Now, pay attention . . . it's very, very important that the wire lies flat for two reasons: (1) if the wire is twisted, it could lead to an electrical short; and (2) when the top of the connector

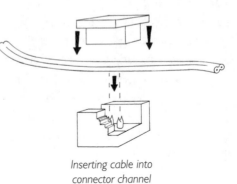

*Inserting cable into
connector channel*

is put back on, it will pierce the wire, which in turn creates the electrical connection. If the wire is on its side, it could miss being pierced.

Repeat the process for all the lights. Cool, huh?

Installing the Transformer

Now you're at the end of the line, so to speak.

In order to connect the end of the cable to the transformer, you'll need to remove some of the insulation. First split the cable in two, either pulling it apart or *carefully* using scissors to cut just the insulation. You just need it to be separated enough to connect the wires to the box. Use the wire stripper to remove enough insulation so that ½ inch of wire is exposed. Place the end of the wire into the proper hole of the wire stripper and the proper length that you want stripped. Clamp down and pull off the insulation.

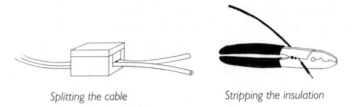

Splitting the cable   *Stripping the insulation*

Some transformers have exposed terminals (a.k.a. screws), and others have terminal screws that are behind a plastic or glass access door, which has to be removed first.

There are 2 terminals and 2 wires, and it doesn't matter which wire gets connected to which terminal. Loosen the terminals with a Phillips or flathead screwdriver, wrap a wire around each terminal, and tighten it with the screwdriver.

Plug the transformer into the exterior GFCI outlet. Follow the manufacturer's directions for adjusting the timer control on the transformer.

Note: Check with the manufacturer's directions to see if sealing the perimeter of the transformer with a silicone sealant is recommended.

Connecting wires to transformer terminals (screws)

Burying Cable

This will be the hardest part of the entire project. And you thought it would be connecting the wires!

You need to bury the cable. So, wearing work boots, start at the end farthest from the transformer and use a lawn edger to dig a 6-inch-deep trench alongside the cable. Don't just dig a little, bury the cable, and then dig some more, etc. It works much better to dig the entire trench line at one time. Also, don't throw away the dirt

Digging a trench

and/or sod; instead, place it nearby on a tarp or large garbage bag because you'll be needing it soon.

Once the line has been dug, lay down the cable, insert the lights, and then fill the trench with the dirt and/or sod. Use your boots to stomp down the dirt piles.

Tools Needed

Flashlights

Graph paper

Pencil

GFCI receptacle
(if necessary)

Low-voltage
landscape lighting

Manufacturer's
instructions

Scissors
(if necessary)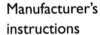

Wire stripper

Phillips or
flathead
screwdriver

Silicone sealant
(if necessary)

Shoes or
work boots

Lawn edger

Tarp or
garbage bag

Replacing a Ceiling Light Fixture

Carmen, a chef, was frustrated at the terrible lighting in her kitchen and swore that replacing it was at the top of her to-do list. But the reality was that she didn't have the dough to buy a new fixture *and* have it installed. And added to that mix was her fear of doing any electrical work. Finally, she mustered up some courage and replaced the fixture herself. Now that Carmen knows how easy electrical projects can be, she swears she'll never put another one on the back burner!

You've hated that ceiling fixture ever since you moved into your house, but thought you were stuck with it because you didn't think you could replace it yourself. Well, this project will be your "lightbulb moment," as Oprah says, because you'll finally see the light as to how easy electrical projects can be, especially this one.

Having said that, there are two caveats: If the ceiling is too high, as with a cathedral ceiling, you may want to hire a certified electrician to do the job. And if you remove the light fixture and you see what's called a "nest"—a big mess of wires—hire a certified electrician to finish the project.

Buying a New Light Fixture

Be sure to measure the base of the existing light fixture, because if you buy one that is smaller, part of the hole in the ceiling may be exposed, which means that you'll probably have to patch it (see pg. 127).

Getting Started

If you're removing a chandelier, you'll need to enlist the aid of a helpful friend. Chandeliers can be heavy and therefore will require an extra set of hands once the wires have been detached.

Open the box with the new light fixture. Read the manufacturer's instructions and check that all the parts are included.

One last thing: don't do this project at night, because the lights will be off in that room and you won't be able to see!

Turning Off the Electricity

Flip the wall switch to the on position for the ceiling light. Turn off the power for the ceiling light at the main service panel. Once you see that the light is off, flip the wall switch to the off position.

Note: If your home's circuit breaker box is not properly labeled, be sure to turn off all computers and televisions in the house before flipping any breakers or removing any fuses.

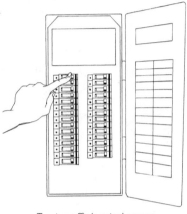

Turning off electrical power

If you have a helpful friend assisting you with this project, you'll need to use two ladders—one for her and one for you. Place the ladders near the light so that you won't be overreaching.

Climb the ladder, wearing safety glasses and the tool belt with the necessary tools inside.

With your fingers, loosen the set screws, nuts, or decorative knob that hold the glass or bracket cover in place. Remove the cover and the lightbulb(s) behind it and hand them to your helpful friend.

Set screws

Loosening set screws or decorative knob

Place the tip of the voltage tester onto the exposed wires. If the voltage indicator emits a continuous beeping sound and/or flashing light, then the electricity was *not* turned off. Go back to the main service panel and try again.

Removing the Old Light Fixture

Take a look at how the wires are connected—they should be black to black, white to white, and green or bare copper to green or bare copper. But that's not always the case. The rule of thumb is that since the wiring works, you'll have to reattach the wires from the new light fixture to the house wires the exact same way. So, draw on a piece of paper exactly how the wires are connected.

Supporting the light fixture, remove the wire nuts and untwist the wires. Hand the fixture to your helpful friend.

Verifying electricity is off

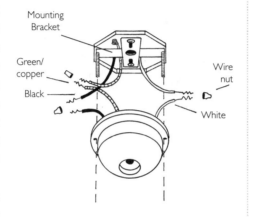

Mounting Bracket

Green/copper

Black

Wire nut

White

Noting wire connection

O f the old light fixture works, don't throw it out. Instead, recycle by donating it to a thrift store or a Habitat for Humanity ReStore in your area.

Installing a New Light Fixture

Almost every new light fixture will fit onto the old mounting bracket. See if the new light fixture will fit over the old mounting bracket screws. If not, install the new mounting bracket by inserting the screws with the Phillips screwdriver.

The wires on the new light fixture should have ½ inch of wire exposed. If there isn't enough wire exposed, then use the wire stripper to remove some of the insulation, if necessary. Place the end of the wire into the proper hole of the wire stripper and the proper length that you want stripped. Clamp down and pull off the insulation.

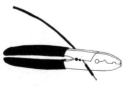

Stripping the insulation

With your helpful friend supporting the new light fixture, twist the green (ground) wires together (one from the new light fixture and the other from the ceiling electrical box) and secure with a new wire nut. Finger-tighten. Twist the white wires together (one from the new light fixture and the other from the ceiling electrical box) and secure with a new wire nut. Finger-tighten. Now twist the black wires together (one from the new light fixture and the other from the ceiling electrical box) and secure with a new wire nut. Finger-tighten. Carefully push the wires into the electrical box.

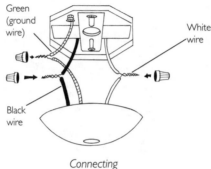

Green (ground wire)

White wire

Black wire

Connecting corresponding wires

Secure the new light fixture in place, following the manufacturer's instructions. Remember to use a CFL bulb with the recommended wattage.

Turn the power on at the main service panel and flip the wall switch to the on position. *Ta dah!*

Securing new light fixture

Tools Needed

Tape measure

New light fixture

Manufacturer's instructions

Helpful friend

Ladders (2)

Safety glasses

Tool belt

Voltage tester

Pencil

Paper

Phillips screwdriver

Flathead screwdriver (if necessary)

Wire stripper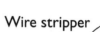

Needlenose pliers

CFL bulb(s)

Replacing a Ceiling Fan (with and without lights)

The ceiling fan in Regina's bedroom was getting about as much use as the exercise bike in the corner. It seemed that when she turned the fan on, it would make a horrible noise, so she'd quickly turn it off. Too bad Regina couldn't turn it into a clothes rack, too.

As the owner of a ceiling fan, you already know the difference it can make in achieving the right temperature for a room all year round. What you might not know is the big difference it can have on your utility bill. In fact, the American Lighting Association states that ceiling fans can reduce an energy bill up to 40 percent in the summer and 10 percent in the winter.

If your ceiling fan is not working adequately, don't assume that you need to replace it. The solution could be one of a number of easy fixes to the fan blades.

Changing the Direction of the Fan Blades

If you feel that the fan isn't circulating the air properly, it could be that the ceiling fan blades are rotating in the wrong direction for the season.

In warm weather, the blades should rotate in a counterclockwise direction to draw hot air upward. In cool weather, the blades should rotate clockwise to force the warm air downward.

Some ceiling fan remote controls have a button to switch the direction, but if not, you'll need to do it manually.

Turn the ceiling fan on to watch which direction the blades rotate. Remember, clockwise for cool weather, and counterclockwise for warm weather. Turn the ceiling fan off.

If you don't have a remote to change the direction of the blades, use a stepladder or ladder to reach the still ceiling fan. Move the slide switch on the motor head of the fan to the opposite direction.

Climb down the ladder and turn the ceiling fan on.

Switching direction of ceiling fan

Tool Needed

Stepladder or ladder

Cleaning the Fan Blades

A rolling stone may not gather moss, but a ceiling fan can gather lots of dirt and dust, even when it's circulating. In fact, if you've never cleaned the top of the fan blades, you'll be in for a real shocker.

It's important to keep the fan blades clean, not only for health reasons but also because too much dirt and dust on a blade can alter the movement of the fan, causing it to be off balance.

There are products specifically designed to clean the blades without having to get on a ladder. Just be sure to turn the fan off before cleaning!

Tool Needed

Extension duster

Balancing the Fan Blades

If the ceiling fan is wobbly, the reason may be that the blades, which come factory-balanced, have become unbalanced over time.

The first thing to check is that the screws are tight. Turn off the ceiling fan and place the ladder near the fan. Wearing the tool belt with the remote control, Phillips screwdriver, painter's tape, pencil, and balancing kit inside, climb up the ladder. Use the screwdriver to tighten all visible screws, being careful not to overtighten the screws on the blades to avoid cracking the wood.

Climb down the ladder (the balance is easier to discern from the floor) and turn on the fan. If it's still wobbly, then turn the fan off and go back up the ladder. This time you're going to remove two adjacent fan blades with the screwdriver and switch their positions. Go down the ladder and turn on the fan. If this hasn't fixed the problem, then you'll need to use the balancing kit. Don't worry—it's super easy.

It's difficult to tell which blade is off balance, so you'll need to identify each blade with a number and test all of them.

Turn off the fan, climb up the ladder, and put a piece of tape on a blade and write "1" on it. Then place a piece of tape on the next blade and mark "2" on it, and repeat on the other blades with "3," "4," "5," etc.

Now place the balancing clip on the outer edge of the first blade, centering it in the middle. Go down the ladder, turn the fan on, and check for any wobbling. If the fan stopped wobbling, then you can adhere a weight (see below). If not, repeat the procedure with the remaining blades, making note of which blade (#1, #2, etc.) wobbled the *least* when it had the clip on it.

With the fan off, place the clip on the edge of the wobbly blade and then move the clip a bit toward the fan body. Step down, turn the fan on, and see if that did the job. If not, keep moving the clip in small increments until you find the right spot. Then remove the backing of one of the weights and

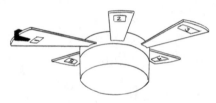

Balancing the fan blades

adhere it to the top of the blade (facing the ceiling), in the center, across from the clip. Remove the clip and tape on the blades.

You may need to repeat the process again and place a weight on another blade. If none of that works, then you'll need to replace the fan.

Tools Needed

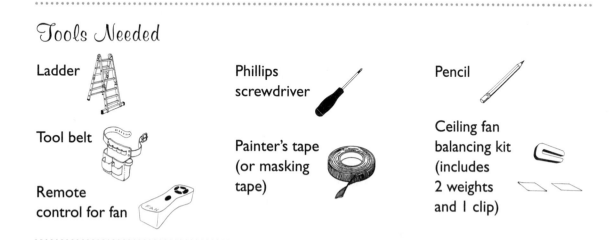

Ladder

Tool belt

Remote control for fan

Phillips screwdriver

Painter's tape (or masking tape)

Pencil

Ceiling fan balancing kit (includes 2 weights and 1 clip)

Buying a Ceiling Fan

With ceiling fans, the sky's the limit when it comes to the wide variety of styles—from a fan that resembles an airplane to one that reminds you of Key West, ceiling fans are now an integral part of a room's decor.

But don't let design alone dictate your purchase. The more important determining factors should be the size of the room, the size of the blades, whether the fan operates by remote control or pull chain, the motor, the mount, the warranty, wet/damp rating (if necessary), and that it has an Energy Star approval. All of these features should be listed on the exterior of the box, and if not, ask a store employee for the information.

And even if the original ceiling fan didn't have lights, you can replace it with one with lights very easily. We promise!

Room Size

It's important to purchase the right size ceiling fan for a room in order for it to be both effective *and* energy efficient.

Ceiling fans are measured by the span of the blades and range in sizes from 29 inches to 84 inches, with 52 inches being the most common.

Don't assume that the ceiling fan you're replacing is the correct size. In fact, don't even bother measuring it. Instead, measure the room.

The American Lighting Association recommends the following guidelines to purchase the right size fan:

Room Size	Fan Size
Up to 75 sq. ft.	29–36"
76–144 sq. ft.	36–42"
144–225 sq. ft.	44"
225–400 sq. ft.	50–54"

Blades

There are two things to consider when it comes to the fan's blades: (1) pitch, and (2) material and finish.

Pitch is the angle of the fan's blades. The greater the pitch, the greater the amount of air that the blades will move. Think of it like putting an oar in the water—the right angle will move more water. An optimal pitch for a fan blade is anywhere from 12 to 15 degrees.

Look for blades that are made of wood or acrylic and are treated to prevent warping, peeling, and scratching.

Remote Control or Pull Chain

A remote control for a ceiling fan should be added to the list of remotes you can't live without. It makes it so easy to change the direction of the blades, as well as the speed, especially if the fan is mounted to a high ceiling. Another reason to choose a remote-controlled ceiling fan over one with a pull chain is that some people pull too hard on the chain, which over time can weaken the fan's support structure.

Motor

Of all the parts of a ceiling fan, the motor is the most important. You can typically tell if the motor is of good quality by the higher cost of the fan ($15 vs. $150), the length of the warranty (five years vs. fifty years), and if it contains sealed bearings, which are permanently lubricated (providing quiet operation).

Mount

The ideal height for a ceiling fan is 8 to 9 feet above the floor. But because houses come in different shapes and sizes, there are mounts to fit all needs.

There are four different types of mounts: (1) *flush*, (2) *standard*, (3) *extended*, and (4) *sloped*. A *flush* mount works best if the ceiling is lower than 8 feet and you want to install a fan with a light. Just be aware that because the blades are so close to the ceiling, the air movement will not be as great as with the other types of mounts. A *standard* mount comes with a downrod that's 3 to 5 inches long. An *extended* mount works well for high ceilings because the downrod can come in lengths from 6 inches (best for 9-foot ceilings) to 120 inches (best for 20-foot ceilings). A *sloped* mount is perfect for ceilings that are vaulted or angled.

Warranty

Get a fan with the best warranty, because the better the warranty the better the fan. Casablanca offers an incredible warranty for the first 120 days of ownership. It will actually send a trained representative to your home to fix any problems. That's in addition to its lifetime motor warranty. Now you know why we chose Casablanca fans for this project.

Wet/Damp Rating

If you've wondered why the fan on the back porch just didn't hold up well, it could be that the original model didn't have a UL wet rating. The wet rating certifies that the motor and housing are moisture resistant, the motor is sealed, the hardware is stainless steel, and all the blades are weatherproofed. If you're replacing a fan that's in a bathroom, then purchase one with a damp rating, for similar reasons.

Energy Star Rating

The Energy Star rating is easy to spot: it's the blue box on the product. This ensures that the appliance has met strict energy efficiency guidelines set by the U.S. Environmental Protection Agency (EPA) and Department of Energy (DOE).

Now that you've purchased the right ceiling fan for your room, let's get started.

Getting Started

Since you'll have the electricity off in the room, try to do this project on a sunny day so you can use the natural light.

Be sure to remove the items from the box, check that all of the parts are included, and read the manufacturer's instructions.

Get all the tools, fasteners, and parts lined up in order of use because once you start this project, you won't be able to stop and search.

Last but not least, we strongly recommend that you have a helpful friend assist you with this project to hand you tools while you're on the ladder and to help you hold up the fan, too.

Safety First

If you have a helpful friend assisting you with this project, you may need to use two ladders—one for her and one for you. Place the ladder(s) near the fan so that you won't be overreaching.

Turn on the ceiling fan. Turn off the power for the ceiling fan at the main service panel. Once you see that the ceiling fan has stopped, you can continue (if the ceiling fan is operated by a wall switch, turn it to the off position).

Note: If your home's circuit breaker box is not properly labeled, be sure

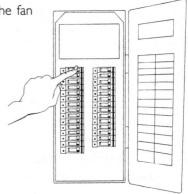

Turning off electrical power

to turn off all computers and televisions in the house before flipping any breakers or removing any fuses.

Removing the Old Fan

Climb up the ladder, wearing the tool belt with the Phillips screwdriver, voltage tester, pen, paper, and needlenose pliers inside.

Removing the Ceiling Fan's Lights

If the old ceiling fan has lights, remove the light bulbs and hand them to your friend. Using your fingers, loosen the set screws (typically 3) on the sides of the glass covers, and remove the covers. Hand these off, too.

Don't be intimidated by the next steps. It may be a lot of information, but it's really easy to do!

Now that the light fixtures have been removed, you need to remove the light housing, which is located under the fan blades.

Have your friend climb up her ladder so that she can support the light housing while you take it off. Use the Phillips screwdriver to remove the screws that connect the light housing to the body of the fan. Slip the housing down to expose the wiring.

Place the tip of the voltage tester onto the exposed wires. If the voltage indicator emits a continuous beeping sound and/or flashing light, then the electricity was *not* turned off. Go back to the main service panel and try again.

You'll see that the wires are connected with wire nuts or there is a modular connection (a plastic rectangular-shaped connector) or both. If there are wires, remove the wire nuts, and untwist the wires. If there's a modular connector, look at it carefully because there's usually a tab that is either pushed in or pulled out to separate the two ends. Now you can remove the light housing.

Removing the light housing

Removing the Fan Blades and Blade Irons

Take a good look at the ceiling fan. You'll see that each fan blade is attached to a blade iron, which is attached to the fan motor. Use the

Phillips screwdriver to remove the screws that attach the blade iron to the fan. Save the screws for possible future use and place the fan blade/ blade iron assembly inside the new box. Repeat with the other blade irons.

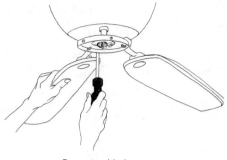

Removing blade iron from fan

Removing the Cover Plate

On the ceiling, you'll see the cover plate, which conceals the hole in the ceiling where the electrical wires are located. Use the Phillips screwdriver to remove the screws that attach the ceiling plate to the electrical box, and slide it down. *Don't worry . . . the ceiling fan is still attached!*

Exposing the Wires

The electrical wires that are tucked inside the electrical box need to be pulled out so that you can remove the wire nuts and separate the wires. Sometimes they're so tightly packed inside the electrical box that you'll need to use needlenose pliers to carefully pull them out.

Now that the wires are exposed, take a look at how they're connected—they should be black to black, white to white, and green or bare copper to green or bare copper. But that's not always the case. The rule of thumb is that since the wiring works, you'll have to reattach the wires from the new light fixture to the house wires the exact same way. So, draw on a piece of paper exactly how the wires are connected.

Removing the Wire Nuts

Have your helpful friend hold the base of the fan.

Unscrew the wire nuts on each connection (there should be 3) and untwist the wires.

Removing the Fan Body

There's still one more thing that's keeping the fan connected—either the fan is attached to a hook or it has a ball-and-socket connection. If there's a hook, simply lift the fan up to release it from the loop. If there's a ball and socket, tip the fan so that the assembly can move through the opening.

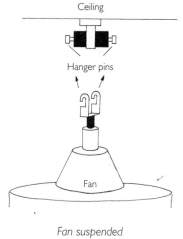

Fan suspended
from hook

Now the fan should be completely disconnected. Hand it to your helpful friend to put inside the new box.

Use the Phillips screwdriver to remove the screws on the mounting bracket, and take the bracket off.

Note: If the electrical box is not specifically made for a ceiling fan (it should say "Approved for ceiling fan mounting" on it) or it's plastic, then it will need to be replaced. You can purchase a ceiling fan electrical box kit at a home improvement store. Follow the manufacturer's instructions for installation.

Climb down the ladder.

Installing a New Ceiling Fan

Note: The anatomy of every ceiling fan is basically the same, but the manufacturers' terminology may differ.

Getting Started

You'll notice that the wires attached to the new ceiling fan are really long, but don't panic. The manufacturer gives you a lot more wiring than you'll actually need. Use the wire stripper to cut the wires to the desired length. Then strip the end of each wire so that about ½ inch is exposed. Place the end of each wire into

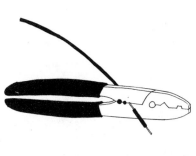

Stripping wires

the proper hole of the wire stripper and the proper length that you want stripped. Clamp down and pull off the insulation.

Read over the manufacturer's instructions as well as ours. You'll see that installing a ceiling fan is really just doing everything in the reverse order of removing a ceiling fan.

Also, remember to have all the necessary parts handy and in order of need.

Installing the New Mounting Bracket

Climb up the ladder, wearing the tool belt with necessary parts and tools inside.

Before you install the new mounting bracket, you need to push the wires from the electrical box through its center hole. Now align the mounting bracket so that the other holes fit over the electrical box screws. Affix the mounting hardware with your fingers and then use the ratchet to tighten.

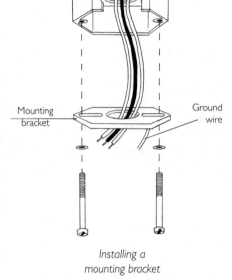

Installing a mounting bracket

Installing the Canopy and Downrod

Have your friend hand you the canopy, and attach it by inserting the long canopy screws using the Phillips screwdriver, being careful not to overtighten.

Now you'll need the downrod. Push the wires from the electrical box completely through the down-

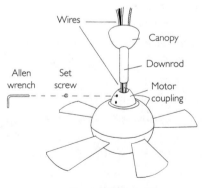

Threading downrod into motor coupling

rod. You may need to loosen the top set screws to allow for its installation. Thread the downrod into the motor coupling, making sure that it's tight (remember, *righty tighty, lefty loosey*). It will require several full turns to lock in place. Insert the set screw to lock the downrod in its position and tighten with the Allen wrench.

Now you'll need to bring the fan up the ladder, so if it's too heavy, have your helpful friend assist you—not on your ladder, of course.

Slide the ball attached to the downrod assembly into the canopy opening so that it's resting properly. Once that's done, it's okay to let go of the fan.

You're almost done!

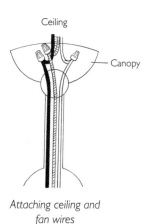

N ever carry the fan by its wires.

Connecting the Wires

Now you're going to attach the wires from the electrical box (which you pulled through the downrod) with the wires from the new ceiling fan. *Remember to refer to your notes regarding the original wiring.* Starting with the 2 grounding wires, twist them together and screw a wire nut on top. Finger-tighten. Next, twist the 2 white wires together and screw a new wire nut on top. Finger-tighten. Repeat these steps with the 2 black wires.

Ceiling

Canopy

Attaching ceiling and fan wires

When you push the wires back into the electrical box, you need to have the wire nuts pointing up and the white and black wires on opposite sides of the canopy.

Installing the Canopy Hatch

Install the canopy hatch by inserting the screws with the Phillips screwdriver (you may have to tilt the fan body a bit in order to get access to the screws). Afterward, check that there's no movement between the canopy and the ceiling. Climb down the ladder.

Removing the Shipping Blocks

Shipping blocks are temporarily installed by the manufacturer to protect the motor during shipping. *These must be removed before you install the fan blades!*

If the blocks are secured with screws or bolts, use a Phillips screwdriver to remove them but don't throw them out—save them in a plastic baggie. Some manufacturers will require you to use those screws or bolts to insert into the fan blades and into the motor.

Attaching the Fan Blades

Some ceiling fan blades have a different color on each side, so be sure to not only pick the color you want but also be sure that each blade will be showing that color.

Attach a blade iron to a fan blade using the screws (typically 3) and the Phillips screwdriver. Be careful not to overtighten, because you may crack the blade. Repeat on all the blades.

Climb up the ladder and have your friend hand you a fan blade. If necessary, have your friend hold the fan blade at the correct spot on the fan motor so that you can insert the screws with the Phillips screwdriver. Repeat with the other fan blades.

WHAT TO DO WITH A BROKEN OR LOST REMOTE

This pertains to ceiling fans that are operated by a remote control and not a pull chain. If the ceiling fan is on and the remote control is broken and the battery is not the problem, here's what you should do: turn off the circuit breaker for the ceiling fan at the main service panel. Wait one minute and then flip the circuit breaker back on. This, of course, shuts off the power to the ceiling fan, but it also reprograms the fan so that it will turn on again only by using the remote control. Contact the ceiling fan manufacturer for a replacement remote, and provide the make and model number of the fan.

Installing the Lights

Install the lower canopy to the motor housing, following the manufacturer's instructions. Attach the globes to the light fixture and twist in the light bulbs.

Note: If you're installing halogen bulbs, don't touch them with your fingers, because the oils from your skin can shorten the life of the bulb. Instead, use a soft cloth or gloves. But in case you did, you can solve the problem by gently wiping the bulb with some rubbing/isopropyl alcohol and a soft cloth.

Turn on the power at the main service panel. Turn the fan on and enjoy the breeze!

Tools Needed

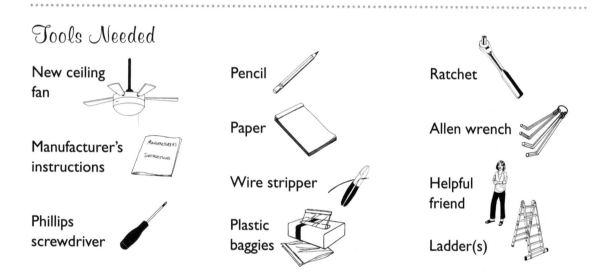

New ceiling fan

Manufacturer's instructions

Phillips screwdriver

Pencil

Paper

Wire stripper

Plastic baggies

Ratchet

Allen wrench

Helpful friend

Ladder(s)

Installing a Ground Fault Circuit Interrupter (GFCI)

When Jeryl was baby-proofing the electrical outlets around her house from her crawling daughter, she never thought about safe-proofing the ones at the kitchen and bathroom sinks for herself. It seems that those outlets were never updated to GFCIs, putting her at great risk. Jeryl was shocked at the discovery, but at least not literally.

You know the funny-looking receptacle that you always see in hotel bathrooms? Well, you *should* be seeing one in your kitchen, in every bathroom, and even outside your home. It's called a ground fault circuit interrupter (GFCI), and it's designed to shut off the electricity emanating from a receptacle (a.k.a. outlet) before you get shocked or electrocuted. In other words, its job is to protect us from ourselves!

A GFCI works by detecting leaks in the electrical current in a circuit branch. If a GFCI discovers even a minute difference between the amount of current flowing into and out of the receptacle, it immediately cuts off the power to it.

Homes built since the 1970s should have them installed near the sinks in bathrooms and kitchens. But we know from our own experiences that it's not always the case. With a few tools and very little time, you can do what the builder *should* have done and make your home much safer.

Once you've done this project in your home, go to your parents' house and install GFCIs there, too. This is a great way to ensure your parents' safety.

Replace any exterior receptacles with GFCIs.

Getting Started

Turn off the power to the receptacle at the main service panel. Use the voltage tester to check all of the receptacle slots. The reason for checking all of the slots is because the receptacle may not have been properly wired, so it's better to be cautious. If the voltage tester emits a continuous beeping sound and/or flashing light, then the electricity was *not* turned off. Go back to the main service panel and try again.

Turning off electrical power

Note: If your home's circuit breaker box is not properly labeled, be sure to turn off all computers and televisions in the house before flipping any breakers or removing any fuses.

Using the flathead screwdriver, remove the screw that secures the receptacle cover and then the cover, and place them in a baggie for possible future use.

Checking receptacle slots with voltage indicator

Removing the Receptacle

Remove the screws on the top and bottom that hold the receptacle to the box in the wall. Gently pull the receptacle out.

The wires are either wrapped around terminals (a.k.a. screws) on the sides of the receptacle, or they're inserted into

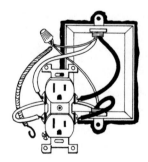

Pulling out the receptacle

holes in the back of it. If wires are attached around the side terminals, loosen the terminals with a screwdriver and detach the wires.

If the wires are inserted in holes in the back of the receptacle, place the small flathead screwdriver in the slot located next to each hole to release the wire. If this doesn't work, use the wire stripper to snip off the wire as close to the receptacle as possible.

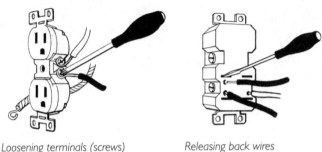

Loosening terminals (screws)   Releasing back wires

Stripping the Wires

In order to connect the wires to the terminals, you will need to have about ½ inch of the wires exposed. Use the wire stripper to snip off the necessary amount of insulation. Place the end of the wire into the proper hole of the wire stripper and the proper length that you want stripped. Clamp down and pull off the insulation.

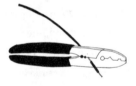

Stripping insulation
from a wire

Installing the GFCI

You'll notice that there are three different colored screws on the GFCI: green=grounding wire; silver=white wire; and brass=black wire. You'll also notice that all the screws are loose. That's because you're going to wrap the coordinating wires around them and then tighten each one. So, here we go . . .

Using the needlenose pliers, wrap the green wire around the green terminal clockwise, about ⅔ of the way around. Use the Phillips screwdriver to tighten the terminal. Next, wrap the white wire around the silver terminal clockwise, about ⅔ of the way around.

Use the Phillips screwdriver to tighten. And finally, wrap the black wire around the brass terminal, about ⅔ of the way around. Use the Phillips screwdriver to tighten.

Note: It may take some finessing, but just stick with it.

When all of the wires have been properly connected, carefully push the GFCI receptacle back into the electrical box, making sure that the grounding wire is not touching the other wires (you may need to use the needlenose pliers to gently move it away).

Use the Phillips screwdriver to insert the long screws that will connect the GFCI receptacle to the electrical box. Use the smaller screw to attach the cover, being careful not to overtighten to avoid cracking the cover, if it's plastic.

Restoring Power

Turn on the power to the GFCI receptacle at the main service panel. Plug a lamp into the receptacle and turn it on to make sure that the electricity has been restored. To test that you installed the GFCI correctly, push the "test" button. This should cause the light to go off and the "reset" button to pop out. To restore the electricity, just press the "reset" button.

To be sure that the GFCIs are operating properly, depress the "test" and then the "reset" buttons once a month on all of them in and around your home.

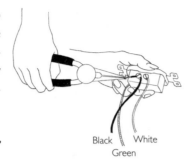

Black   White
Green

Wrapping coordinating wires around colored terminals

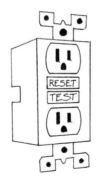

Testing GFCI

Remember to install the **GFCI** right side up, which means that the holes should resemble a face.

Tools Needed

GFCI receptacle(s)

Voltage tester

Small flathead screwdriver

Plastic baggie

Wire stripper

Needlenose pliers

Phillips screwdriver

Plug-in lamp

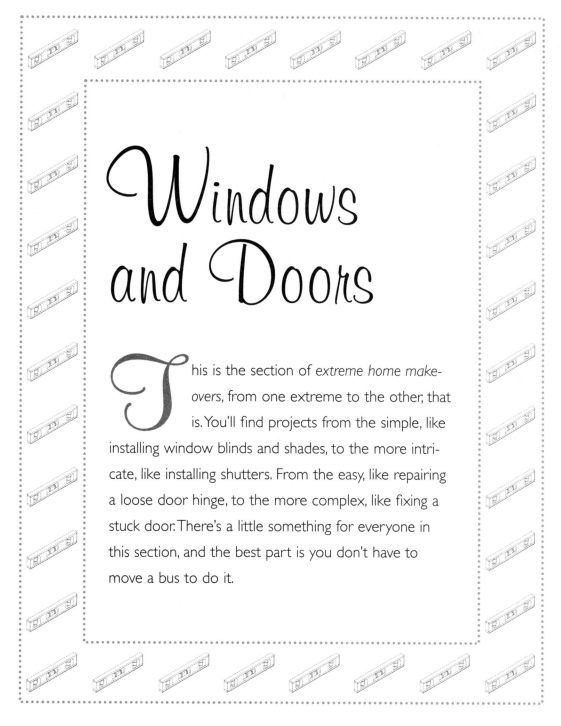

Windows and Doors

This is the section of *extreme home makeovers*, from one extreme to the other, that is. You'll find projects from the simple, like installing window blinds and shades, to the more intricate, like installing shutters. From the easy, like repairing a loose door hinge, to the more complex, like fixing a stuck door. There's a little something for everyone in this section, and the best part is you don't have to move a bus to do it.

Installing Window Blinds, Shades, and Plantation Shutters

*A*re you still living like a college student . . . three jobs later? If so, it's time to take the sheets off the windows and graduate to real window coverings, like blinds, shades, and shutters. The project is elementary!

Today's window blinds, shades, and shutters do so much more than provide privacy—some are even designed to block unwanted ultraviolet rays and insulate windows, too. And with such a wide variety of colors, styles, and prices, there's one to fit every window . . . and wallet.

Buying Window Blinds and Shades

When deciding whether to dress your windows with blinds or shades, don't just think about the look—think about the function, too. Blinds are the optimal choice for providing privacy while at the same time allowing light to enter. But if you want a room to be darkened, consider buying a "blackout" shade, which can actually block out about 99 percent of exterior light.

If you need the window treatment to provide insulation, look for shades with backing, or double-layer cellular shades. These shades have what's called an R-value, which is the quality rating for its ability to insulate. The higher the number, the more energy efficient the shade.

Wood blinds are also good for insulating a window. Just be sure not to install them in a high-moisture area, like a bathroom.

Depending on the size of the windows and your preference for material and style, you may have to special-order them, or you may be able to walk into a home improvement store and have them cut to size while you wait.

But no matter where you buy window treatments, it's up to *you* to get the measurements right. Once the window covering is cut to size, it cannot be returned. So, *measure twice* so you don't *pay twice!*

Getting Started

We strongly recommend that you purchase a tape measure that's easy to read and that has ⅛ inch markings clearly defined. *Every fraction counts when it comes to measuring!*

But before you take out the tape measure, you need to understand the two mounting options: (1) an inside mount (inside the window opening), which is the most common, is designed to provide a neat fit; and (2) an outside mount (on the wall or trim), which is used for added privacy or if there is insufficient window trim. We're going to show you how to do both.

Don't assume that your windows are perfectly square. They're not. And don't assume that every window in your home has the same measurements. They don't. Therefore, measure *every* window that needs a window covering.

Measuring for Interior Mounted Blinds and Shades

Note: If you're measuring for replacement blinds, you need to measure the window and not the old blind.

To get the most accurate measurements for the width, you'll need to measure the inside frame of the window, from the left to the right, in 3 places: top, middle, and bottom. Write down the measurements, rounding to the nearest ⅛ inch. *Remember, the smallest of fractions is important!*

Now repeat the process for accuracy. For the *width*, use the *smallest* of the 3 measurements.

Taking width measurements

To get an accurate measurement for the height, measure from the windowsill to the flat spot in trim of the window frame, in 3 places: left, middle, and right. Write down the measurements, rounding to the nearest ⅛ inch. *Remember, the smallest of fractions is important!*

Now repeat the process for accuracy. For the *height,* use the *largest* of the 3 measurements.

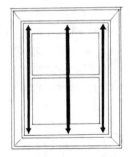

Taking height measurements

Measuring for Outside-Mounted Blinds and Shades

As previously stated, this mounting option is typically used for additional privacy, so the width and height is going to be determined by how much you want the blind to extend over the window. Typically the width is 4 to 6 inches more than the window opening.

The height measurement for outside-mounted *blinds* is typically 3 to 6 inches more than the length of the window. For *shades,* 2 to 4 inches is typically added for the length.

Measure in 3 places: left, middle, and right. Be sure to be accurate with your measurements, rounded to the nearest ⅛ inch. Use the largest of the 3 measurements.

Window Blinds

Finding the Location for the Brackets for an Outside Mount

There are two different mounting brackets for a window blind: one for the right and one for the left. The problem is that they're *never* marked, and every bracket is designed a little differently by the manufacturers. So, the best way to know which is which is to refer to the directions. And if you're still not sure (because manufacturers' directions can be pretty bad), call the company's 800 number.

Set up a ladder or stepladder near the window, using the appropriate safety measures.

Measure the width of the window and mark the center above the window trim. Measure the width of the headrail (the top part that gets mounted) and mark its center with a pencil. Then line up the marks and it should be centered.

Once the head-rail has been centered on the wall, use a pencil to run a light line the entire width of it, along the wall.

Remove the blind and place the level on the line to see if the bubble in the small window is between the two lines. If yes, then the line is level. If not, adjust the pencil line.

Now add ¼ inch to both sides of the line and mark with a pencil. This is where you'll install the brackets.

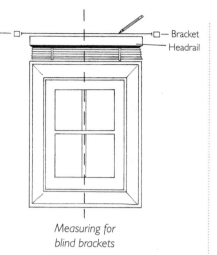

Measuring for blind brackets

Installing the Brackets

Place a bracket on the mark and pencil in the screw holes. Repeat with the other bracket. Wearing the safety glasses, use the drill to create starter holes in those penciled spots. Position the appropriate bracket so that it's aligned over the holes. Use a Phillips screwdriver to insert the screws.

Tools Needed

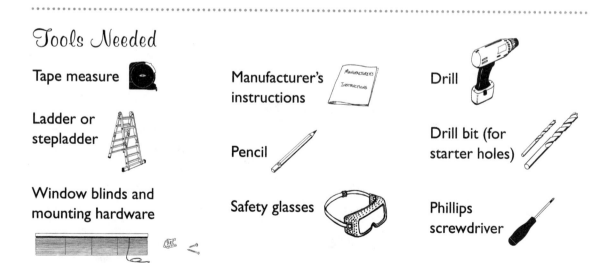

Tape measure

Ladder or stepladder

Window blinds and mounting hardware

Manufacturer's instructions

Pencil

Safety glasses

Drill

Drill bit (for starter holes)

Phillips screwdriver

Installing the Blind

Snap the headrail into the brackets, with the cord on the right side. This is typical for all standard blinds. If you're ordering custom-made blinds, you can choose the location of the cord. Follow the manufacturer's directions regarding cord safety.

Window Shades

Installing Brackets for Interior Mount

The two mounting brackets have different openings—one is a circle hole and the other is a slot. The bracket with the circle hole should always be mounted on the right side of the window. The bracket with the slotted opening should always be mounted on the left.

Installing interior mount
window shade brackets

Tools Needed

Ladder or stepladder

Window shade and mounting hardware

Pencil

Tape measure

Safety glasses

Drill

Manufacturer's instructions

Drill bit (smaller than the mounting screws)

Phillips screwdriver

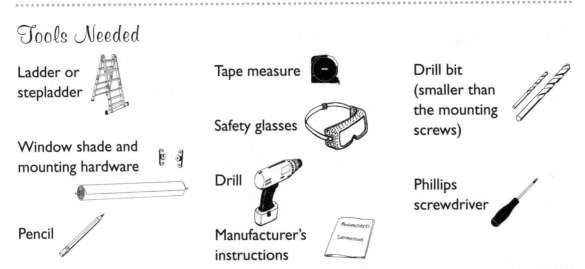

Set up a ladder or stepladder near the window, using the appropriate safety measures.

Hold the shade up to determine the best location, making sure to leave enough space at the top for it to roll and unroll. Use a pencil to mark on the right side of the window where the round end pin hits the trim.

Measure the distance from the mark to the top of the inside of the window. Use this measurement to mark a corresponding location on the left side for the other bracket.

Hold the shade so that the round end pin and the other end are on their marks. Adjust the shade's marks, if necessary.

Position the circle hole of the bracket over the mark on the right side. While holding the bracket with one hand, use the other hand to pencil in the screw holes onto the trim.

Wearing safety glasses, use the drill to create pilot holes. Then place the bracket over the holes and use the Phillips screwdriver to insert the screws.

Repeat the procedure on the left side.

Installing the Shade

To install the shade, push the round end pin, which retracts, into the bracket with the circle hole, and then slide the other end into the slot.

Buying Plantation Shutters

You can't help but think of the Old South when you see plantation shutters. Today, plantation shutters serve as more than just a way to keep your privacy while allowing air to circulate . . . they're also considered "window furniture." In fact, plantation shutters are the only window covering that adds financial value to a home.

Just like blinds and shades, shutters come in a variety of colors and can even be custom colored. In addition, shutters are made of synthetics or wood, with synthetics being the optimal choice for high-moisture areas.

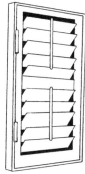

Plantation shutters

Getting Started

First, you need to decide what type of mount you want for the shutters—inside or outside. Unlike with blinds and shades, an exterior mount is more popular than the inside mount. When choosing the mount, you need to take into account if the window has enough framing width to support an inside mount. Also, if the window is a tilt-in style (typically used for easy cleaning), then an inside mount would prohibit that particular movement. Therefore, you would want an outside mount.

Whether you're purchasing shutters online or buying them in a store, it's up to *you* to get the measurements right. Once the window covering is cut to size, it cannot be returned. So, *measure twice* so you don't *pay twice*!

We strongly recommend that you purchase a tape measure that's easy to read and has ⅛ inch markings clearly defined. *Every fraction counts when it comes to measuring!*

One more thing about plantation shutters . . . pay attention to the manufacturer's warranty regarding use and care of the product. These window coverings are expensive, and you don't want to do something that will void the warranty.

Measuring for Inside Mounting of Plantation Shutters

To get the most accurate measurements for the width, you'll need to measure the window from left to right (from the flush flat spot in the trim to the next flush flat spot in the trim) in 3 places: top, middle, and bottom. Write down the measurements. *Remember, the smallest of fractions is important!*

Now repeat the process for accuracy. For the *width*, use the *smallest* of the 3 measurements. You will need exactness to $\frac{1}{16}$ inch.

Note: If you have a casement window with a crank on the bottom or a lock on the side, you must make note of the crank's or lock's exact location and dimensions so that the manufacturer can create cutouts.

To get an accurate measurement for the height, measure from the windowsill to the flat spot in the trim or where you want the shutter to stop (e.g., if you don't want the shutter to cover a transom).

Measuring for Outside Mounting of Plantation Shutters

To get the most accurate measurements for the width, you'll need to measure the window from left to right (from the flush flat spot in the trim to the next flush flat spot in the trim) in 3 places: top, middle, and bottom. Write down the measurements. *Remember, the smallest of fractions is important!*

Now repeat the process for accuracy. For the *width*, use the *smallest* of the 3 measurements.

To get an accurate measurement for the height, measure from the windowsill to the next flat spot or where you want the shutter to stop.

*Measuring window for
an outside mount*

Installing Plantation Shutters

You will be amazed at how fast you'll be able to install these plantation shutters . . . that is, as long as you use the screws we recommend (i.e., box head self-tapping screws)!

Inserting top screw

Inserting screw near hinge

Because of the size of the shutters, you'll need to give yourself enough work room. In fact, you'll want about 2 to 2½ feet, so move any furniture nearby. Set up a ladder or stepladder near the window, using the appropriate safety measures.

Insert the first screw at the top of the window to hold the shutter in place. Then insert a screw near each hinge, and then one at the bottom. The typical plantation shutter will require 8 screws.

Tools Needed

Tape measure

Plantation shutters

Ladder or Stepladder

Eight 2-inch box head self-tapping screws (per window)

Manufacturer's instructions

Drill

Drill bit (for box head self-tapping screws)

Fixing a Loose Door Hinge

*I*f you look at most home repair books and do-it-yourself Web sites, the solution to the problem of a loose door hinge is usually to push either toothpicks or a golf tee into the screw hole. But for those of you who don't pick your teeth or play golf, we have an inexpensive alternative to get the door swinging right.

With the constant opening and closing of a door, screws in a door hinge can become loose over time. Typically, it's because the screws weren't long enough to dig in and hold onto the wood. We can't explain why door installers insist on using short screws instead of long ones—the cost difference is negligible. What we do know for sure is that if you don't address this problem immediately, it can lead to bigger ones, such as damage to the weatherstripping, hardware, and door frame.

Buying Tools

Before going to the store, you need to know how many screws need to be replaced. Open the door and use the Phillips screwdriver to check the tightness of all the screws in the hinges. Remove the loose screw and check to see if the problem is that it's too short (½ inch). If so, then the other screws are probably the same length, which means that you should replace them, too.

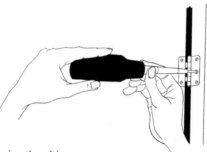

Checking hinge screws

We think the best tool for this job is the one a lot of contractors use, which is a stripped screw hole repair kit. This contains strips of steel mesh, is *very* inexpensive, and can be found in most home improvement stores.

The other important thing to buy is a screw that's longer than the one that's installed in the door hinge. You'll need to remove one screw and take it to a home improvement store to find a replacement of the same width, but about ½ inch longer. Be sure to buy extras.

Getting Started

The beauty of this project is that you don't have to remove the door! Just recheck all the hinges screw to make sure that you didn't miss any that need to be replaced.

Fixing a Loose Door Hinge

Use the Phillips screwdriver to remove the loose screw and throw the screw away. With the scissors, cut a piece of the steel mesh and insert it lengthwise into the hole. You may need to trim it a few times to get just the right size. If it fits perfectly, cut another one to the same size. Push both into the hole opposite each other.

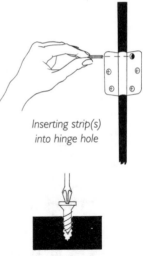

Inserting strip(s) into hinge hole

Using the Phillips screwdriver, insert the new (longer) screw into the hole and tighten.

Repeat the process on any other screws that need to be replaced.

Steel mesh gripping screw

Tools Needed

Phillips screwdriver

Mr. Grip (stripped screw hole repair kit)

Longer screw(s)

Scissors

Fixing a Stuck Door

We wonder, would **Mister Rogers** have sung a different tune if his front door was always jammed? It can be a beautiful day in the neighborhood for you, too, with this simple repair.

You've probably noticed that as the seasons change so does your ability to easily open the front door. Weather, especially heat and humidity, can cause a wood door to swell. But what if you're having problems all year round, the door's not wood, and it's located in a bedroom?

Well, you can't blame Mother Nature for that! What can be faulted are loose hinge screws, too much paint on the door and/or door frame, or the fact that the door was not properly hung.

So, tighten or replace any loose screws (see "Fixing a Loose Door Hinge") and remove any excess paint, if necessary. If the problem still exists, then you'll need to sand the door.

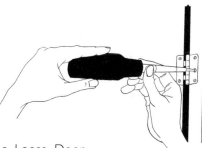

Checking tightness of hinge screws

Getting Started

You can definitely do this project without removing the door if the problem area is at the top or on the edge of the door. If it's at the bottom, then you have to remove it.

You can do this project with a sanding block and sand paper. However, we strongly recommend that you use an electric sander because it will cut the time by a lot. So, buy, borrow, or rent an electric sander, because your time is worth it.

With the door closed, take a look at the opening between the door and the door frame. Can you see where there is less space than anywhere else? If so, that's where the door is getting stuck. Tear off a piece of painter's tape and apply it to the door directly below that location.

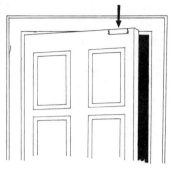

Marking a door

If the door is dragging on the floor, place a piece of tape above the problem area. No matter where the problem is located, you won't need to sand too much.

Fixing the Problem Area at the Top or Edge of the Door

Now that you have the door open, use the painter's tape to adhere the plastic drop cloth behind the door to prevent dust from going into a room or closet. Spread the drop cloth on the floor underneath the door.

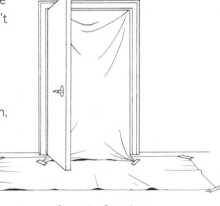

Protecting from dust

Place a stack of shims or magazines under the edge of the door to keep it stationary.

Use scissors to cut the sandpaper to size and attach it to the electric sander. If necessary, plug the sander into an extension cord.

Wearing the safety glasses and dust mask, climb up the stepladder (if the problem area is at the top of

Using shims to keep door stationary

*Sanding with
protective equipment*

the door) and sand the area above where you marked with the painter's tape. Be careful to not sand beyond the taped area. It shouldn't take very long to complete, so after a minute or two, turn off the sander, step off the ladder, remove the shims, and check your work.

If the door still needs more sanding, repeat for a few seconds and then recheck your work. Wipe off the door with the dust cloth. Using a paintbrush, seal the sanded area with an interior polyurethane finish or primer/paint to prevent swelling caused by moisture. Remove the drop cloths.

Fixing the Problem Area at the Bottom of the Door

If the problem area is at the bottom of the door, then you'll have to remove the door.

Remove the shims. Have your helpful friend hold the top edge of the door, opposite the hinges. Starting at the bottom hinge, place the screwdriver underneath the hinge pin and hit it with the hammer. Place the pin on a paper towel or newspaper—it can be greasy. Repeat the process with the remaining hinges until the door is completely removed.

Lay the door on the floor cloth, and wearing the safety glasses and dust mask, sand the problem area you marked with tape, being careful not to sand beyond that area.

Note: Unlike with sanding the top or edge of the door, you won't be able to check your work as you go.

When you're done sanding, wipe off the door with the dust cloth. Using a paintbrush, seal the sanded area with an interior polyurethane finish or primer/paint to prevent swelling caused by moisture. Remove the floor cloth and reinstall the door with the help of your friend.

Tools Needed

Electric sander or sanding block

Sandpaper (100 grit— the higher the number the finer the grit)

Painter's tape

Plastic drop cloth (or plastic tablecloth or shower liner)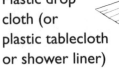

Shims (or a few magazines)

Extension cord (if necessary)

Safety glasses

Dust mask

Stepladder or ladder

Dust cloth

Paintbrush

Interior polyurethane finish or primer/paint

Helpful friend

Phillips screwdriver

Hammer

Paper towels or newspaper

Floor covering

Scissors

Ladder or stepladder

Weatherstripping Windows and Doors

*L*inda was sitting alone in her living room on a cold winter night, curled up with a new thriller, when she noticed that the drapes were stirring although the window was closed. Linda thought this mystery could be the making of a new Nancy Drew book, *The Case of the Moving Curtains.* Where are Bess and George when you need them?

The only crime that was committed here was a lack of weather-stripping.

The U.S. Department of Energy states that in a typical home, about one-third of its energy is lost through its windows and doors. That doesn't mean that you're keeping them open while you're running the heater or air conditioner. What it does mean is that the windows and exterior doors in your home are not properly sealed, and therefore, cooled or heated air is escaping.

The optimal solution would be to replace all the windows and exterior doors with new energy-efficient models. But for the majority of us, that's not an option. And that's why we're providing you with inexpensive and easy ways to keep the costly cooled and heated air (as well as your money) inside your home.

Okay, so you know that weatherstripping for windows and doors is important, but how do you know if your home needs it? One way is to open the front door, put a piece of paper into the jamb and close the door. Try to pull the paper out. If you're successful, then there's an air leak. Now try it with a window. Again, if the paper comes out, then there's an air leak.

The U.S. Department of Energy recommends that you turn down the dial on the thermostat in the winter by 10 to 15 degrees for 8 hours a day to save yourself 5 percent to 15 percent on your home's annual heating costs. The same holds true during the summer by keeping your home more warm than cool.

If you want an exact calculation of how much energy your home is wasting, the problem areas, and how to fix them, hire an energy audit professional. Contact your state or local government energy department or your local utility company to request a list of reputable contractors in your area.

And when you start contacting energy auditors, ask if they use a calibrated blower door and if they perform a thermographic inspection, both of which are very important in getting the most accurate information.

Buying Weatherstripping

The optimal way to purchase is through the window and door manufacturers so that the weatherstripping is a perfect fit. Most windows and doors will have a manufacturer's sticker on them. Or you can get the information from your home builder.

But if that's unrealistic, then do what most homeowners do and purchase the weatherstripping at a home improvement store.

Types of Weatherstripping

Weatherstripping comes in different lengths and widths. To determine the amount of weatherstripping you'll need, measure the perimeter of the doors (remember, exterior doors only) and windows (if all the windows are the same, you have to measure only one and then multiply the measurements times the number of windows in your home). In case you don't remember: perimeter=sum of all the lengths. Be sure to factor in an additional 10 percent for waste.

To determine the width, take off a piece of the old weatherstripping and either measure it or bring it into the store to find a replacement of equal width.

There's a wide variety of weatherstripping: felt, foam, vinyl, rubber, and metal. Felt and foam are the least expensive and are really easy to install, but they don't hold up over time and are not the most energy-efficient. Metal is much more energy-efficient but is more expensive. It's also more difficult to mount, which is why it's typically installed by a contractor. So, that leaves vinyl and rubber.

We chose to use rubber weatherstripping for this project instead of vinyl, because the rubber model had a self-adhesive backing (the vinyl requires using a hammer and nails to install), and it came with a warranty (the vinyl had no warranty).

Now, here's a helpful hint—you don't have to use expensive weatherstripping on all the windows. Use the inexpensive foam (with a self-adhesive backing) on windows that don't get much use or are in rooms that are vacant.

Door Sweeps

A door sweep is designed to close the gap between the bottom of the door and the flooring inside your house. It also works to cover up uneven spacing that occurs with a sagging door. Door sweeps come in different colors, styles, and price ranges, but because the door gets a lot of use, you shouldn't go cheap. Instead, invest in a high-quality product.

Be sure to measure the width of the door and determine the desired height of the sweep before purchasing.

Also, consider adding sweeps to the bottom of the garage door(s). There are products specifically made for these doors.

Weather-stripping typically comes in white, gray, and brown, so buy the color that best matches your windows and doors.

Getting Started

Before you get started, you need to understand where to install the weatherstripping on windows and doors, which requires that you have a grasp of the terminology.

A double-hung window is two windows that move independently of each other and lock together in the middle. A casement window has a crank that opens the window out and in.

On a double-hung window, you'll be installing weatherstripping on the bottom of the sash on the lower window and the top of the sash on the upper window. To better understand, open the lower window and feel underneath the bottom for the channel. Now close that window and lower the upper window, and you'll see the channel on the top of the sash.

With the upper and lower windows closed, look at the stops that are located on the vertical jambs on each side. They run above the lower window up to the top of the window frame. These stops are there to keep the lower window from being raised too far and possibly damaging the upper window frame. So, you can't weatherstrip on top of the stops, especially if you have tilt-out windows!

Instead, you need to install weatherstripping in the vertical jambs. Open the bottom window and you'll see that there are multiple channels. Which one is for weatherstripping? Typically, the middle one, but just to be sure, put your finger into the bottom sash, and you'll feel an opening, which runs over one of the channels. That's the one you want to weatherstrip.

For a casement window, crank it open and look at the window frame that encases the window. The channels that are located in the window frame are where you'll be installing the weatherstripping. On some casement windows, additional weatherstripping can be applied to the window border, attached to the house, into where the window seals shut.

And for a door, you'll weatherstrip all around the jamb inside the door frame.

To make things less confusing, refer to the diagrams for windows and doors while reading the instructions.

Cleaning the Windows and Doors

In order for weatherstripping to adhere properly to the window, remove the old weatherstripping and clean the channels

Double-hung window

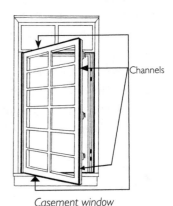

Casement window

Exterior door

in the jambs and sashes using warm, soapy water. For a casement window, clean the channels in the window frame and around the window itself. Let the areas dry completely.

The same holds true for doors. Be sure to remove any old weatherstripping, clean the edges of the door and door frame with warm, soapy water, and let dry.

Read the manufacturer's instructions for applying the weatherstripping. Some products require a room temperature above 40 degrees.

Installing Weatherstripping on a Double-Hung Window

Open the lower window, measure the bottom sash, and cut a piece of weatherstripping about 1 inch *longer*. Peel off the adhesive backing and press it into sash's channel. Either use scissors or a utility knife to remove any extra material. Repeat the process on the top of the upper window.

Pressing weatherstripping into sash channel

Before moving on to the other windows, check your work by opening and closing the window. The goal is to have a good seal with the window still able to move up and down.

Installing Weatherstripping on a Casement Window

Open the window and measure top, bottom, and sides of the window frame. Cut a piece of weatherstripping about 1 inch *longer* than each measurement. Peel off the adhesive backing and press it into the channels of the window frame and window border, if applicable. Use either scissors or a utility knife to remove any extra material.

Applying weatherstripping around casement window and frame

Before moving on to the other windows, check your work by opening and closing the window. The goal is to have a good seal with the window still able to open out and close.

Installing Weatherstripping on a Door Frame

Open the exterior door and cut a piece of weatherstripping about 1 inch *longer* than the doorjamb. It's best to use one long piece for each section and to get a close fit in the corners at the top.

Peel off the adhesive strip on the back of the weatherstripping and push it into place. Use either scissors or a utility knife to remove any extra material.

Applying weatherstripping to a door jamb

Repeat the process on the other two sections of the door frame.

Before moving on to another exterior door, check your work by opening and closing the door. The goal is to have a good seal with the door still able to open and close.

Installing a Door Sweep

The sweep is usually wider than the door, which means that it will need to

Measuring a door sweep

be cut. Place the sweep against the door and draw a line with a marker where the sweep needs to be sliced.

Use scissors to cut through the vinyl. To cut through the metal, you'll need a vise and a hacksaw.

Put a towel on the floor and place the vise on top of it. Before putting the sweep into the vise, wrap a cloth around the sweep to protect it from getting scratched inside the vise. Once the sweep is secured inside the vise, put on the safety glasses and use the hacksaw to cut along the line.

Don't forget to install a sweep on the door that leads to the garage.

Position the sweep against the door so that the bottom part (the vinyl) slightly touches the flooring or threshold. Temporarily secure it to the door with duct tape. Open and close the door to ensure that the sweep is in the correct location. Use a pencil to mark the holes onto the door. Remove the sweep and duct tape.

Cutting a door sweep

Use a drill bit to create pilot holes in the marked areas along the bottom of the door. Position the sweep so that the holes align over the pilot holes. Use the screwdriver to insert the screws into the holes.

Proper placement of a door sweep

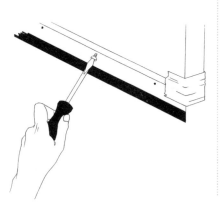

Tools Needed

Tape measure

Weather-stripping

Rag with warm soapy water

Scissors or utility knife

Door sweep

Marking pen

Vise

Hacksaw

Towel

Cloth

Safety glasses

Duct tape

Pencil

Drill bit (for pilot hole)

Drill

Flathead or Phillips screwdriver

Walls

If your walls could talk, would they scream, *"Help"*? And would your floors shout, *"Don't tread on me"*?

In this section, you'll learn how to patch the gaping hole in the ceiling, paint like a pro, and replace the broken floor tile, which has put you in denial for too long. And that's just to get you started.

By the time you're done with a few of these projects, your home will be yelling, *"Look at me!"*

Removing a Glued-On or Oversize Mirror

Every time Lerma looked at the wall-to-wall bathroom mirror, all she could see staring back at her was a must-do project. Lerma finally decided to take a crack at removing the mirror and replaced it with a smaller, more updated model. Now the bathroom mirror reflects something much more beautiful—Lerma's can-do spirit.

There's no need to be intimidated by this project. And more important, there's no need to think that after cracking the mirror you'll have bad luck. Let your success at doing this project be your good fortune!

Getting Started

A mirror is affixed to a wall by using an adhesive, a hook, or mounting brackets. Before you tackle this project, you'll need to know how the mirror is fastened to the wall so that you use the right technique and tools to remove it.

Removing a Glued-On Mirror

If the mirror is attached to the wall with an adhesive, there's no way to avoid damaging the mirror *and* the drywall. What you *can* avoid is doing any harm to yourself. So, even though this method may seem odd, it's the safest technique.

Spread a drop cloth over the countertop or sink and one over the floor. Place a large leaf bag inside the garbage can and move it close to where you'll be working.

If you're disposing of the mirror, whether it's broken or not, contact your waste disposal company to let them know that you are putting it out for pickup. Sometimes the company will notify its employees or ask you to place a handmade sign on the garbage can saying, "Mirror."

Cut the duct tape the width of the mirror and secure it, one strip at a time, starting at the top and working your way down. Be sure there is no glass peeping out from between the strips of tape.

You're going to be removing the mirror in sections, so visually divide the mirror into four sections and pick one area to start.

Note: There will be tiny shards after breaking the mirror, so you must wear work gloves, closed-toe shoes, and safety glasses—prescription glasses and sunglasses will not completely protect your eyes.

Wearing the gloves, safety glasses, and shoes, take the hammer and hit one section of the mirror. Pull up on the corner so that the broken piece is raised. Now use the utility knife to cut the duct tape along the edge of the damaged area. Remove the piece of mirror and place it inside the garbage can. Repeat the procedure until the entire mirror is completely removed.

Cutting away damaged mirror

Tools Needed

Drop cloths (plastic or cloth)

Large paper leaf bag

Large garbage can with liner

Duct tape

Scissors

Work gloves

Closed-toe shoes

Safety glasses

Hammer

Utility knife

Removing an Oversize Mirror

Depending on the size of the mirror, it can be very heavy and awkward to carry—in other words, easy to drop. Therefore, we recommend that you have a helpful friend assist you *and* that you know ahead of time where you're going to place the mirror once you've removed it.

Note: If you have a mirror that runs the length of a room (e.g., a bathroom mirror that the builder put in twenty years ago), you may need to remove it in pieces, even if it's not attached with adhesive, because there may not be enough space to take it down without damaging a wall or even the ceiling.

Note: To prevent any possible injuries, wear work gloves, closed-toe shoes, and safety glasses.

If the oversize mirror is framed, it probably has a wire on the back that hangs on a fastener (screw or nail). To remove the mirror, have your helpful friend stand at one end, and you at the other, and on the count of three, lift it up so that the wire releases from the fastener. Place the mirror in a safe location.

If the mirror is long and unframed, it's probably secured to the wall with brackets and screws that are visibly positioned on the top

*Removing
brackets*

and bottom, running the length of the mirror. To remove the mirror, you and your friend should start at opposite ends. Using the Phillips screwdrivers, remove the top screws, with each of you placing a hand on the mirror so that it doesn't fall forward. You may be able to remove the mirror by loosening just the bottom screws. With the two of you at opposite ends of the mirror, lift it up on the count of three, being careful not to damage any walls, ceiling, or flooring in the process. Place it in a safe location.

Tools Needed

Helpful friend

Work gloves (2 sets)

Closed-toe shoes (2 pairs)

Safety glasses (2 pairs)

Phillips screwdrivers (2)

Hanging Prepasted Wallpaper

After sending her last child off to college, Donna decided it was out with the old and in with the new—wallpaper, that is. Being a novice, she schooled herself on the different types of wall coverings and how to hang them. The project was textbook. Donna was taking the first step toward a large learning curve on how to enjoy her nest, albeit an empty one.

Anyone who's ever hung wallpaper knows it's not as tough as it seams (*hee-hee*). In fact, a lot of people actually prefer hanging wallpaper to painting walls because wallpaper keeps its look longer than paint, so the bragging rights keep going for years.

Wallpapering is definitely something that you can do yourself, but we recommend that you find a helpful friend because four hands work faster than two.

Buying Wallpaper and Products

Measure the room you'll be wallpapering and input the stats (including the number of doors and windows) into a wallpaper Web site to get an estimate of the number of rolls you'll need, so you won't over- or underbuy, and also so that you won't be surprised by the cost.

Wallpaper, which can be purchased in a store or online, is priced per single roll but is sold in single, double, or triple rolls. Wallpaper is also sold in American or European rolls—American is wider and longer—so be careful when you're ordering. And you'll need to take into account if the motif is a "drop match" or a "straight match" (it will say so on the back of the page of wallpaper). A straight match is when you hang one strip of wallpaper next to another and the motif matches side by side. A drop match can waste more wallpaper because it may take more than one strip of wallpaper to find the match.

If you hesitate about wallpapering a room because you dread having to remove it someday, consider York Wallcoverings' new "super strippable" product called *SureStrip*. The beauty of this wall covering is that it can be removed in a single piece, which allows you to change the wallpaper as often as you want.

Some necessary items to purchase for this job (besides the wallpaper) are: seam and repair adhesive (a small tube will be enough), a boxful of single-edge razor blades (you'll be using a lot of them), and a wallpapering kit (typically includes a water tray, a hanger/smoother, and a roller).

Getting Started

You really shouldn't start wallpapering until you've painted the walls with a primer. *Just what you wanted to hear, right?* So give yourself an extra day for this.

Wall covering primer is a paint that allows the wallpaper to adhere better while also making removal easier. If you're going to invest the time and money in covering your walls, then you should invest some time in prepping the walls correctly. Oh, and it's also a good time to paint the ceiling (with a ceiling paint).

Once the walls have been prepped, you can prep the floors by placing drop cloths on the floor where you'll be wallpapering, as well as the room where you'll be setting up shop.

The job will go so much easier if you remove as many things as possible from the walls, such as the towel bars, mirror, toilet paper holder, medicine cabinet, paintings, photos, drapes, etc. And if you're wallpapering a bathroom, you may find it best to remove the tank from the toilet for easier access behind the toilet.

We recommend that you set up a 6-foot worktable, because it enables you to roll out the paper, measure it, and cut it. If you're using a dining room table, be sure to properly protect it.

Determining the Proper Length

The best place to hang the first piece of wallpaper is in the least visible spot in the room. Using a long level or chalk line, create a straight line from ceiling to floor. Now measure the line and add about 2 inches for error. This will be the measurement you'll use for cutting the majority of pieces.

Prepping the Wallpaper

Stretch out a roll of wallpaper on the worktable and, using the measurement from above, mark it and then use scissors to cut the first sheet, or you can do like the professionals and use a straight edge ruler and a utility knife.

Roll the cut sheet in the opposite direction (outside in) to flatten. Roll out more paper on the worktable and lay the cut sheet over it to perfectly match the pattern.

Soaking the Wallpaper

You can cut some pieces ahead of time, but you can hang only one piece at a time.

There are two techniques for properly soaking the wallpaper: (1) using a wallpaper tray, and (2) using a paint roller brush. The most common method is to roll up a piece of wallpaper (with the print facing inside), place it into a water-filled wallpaper tray, and allow it to soak for about 5 minutes.

The method preferred by professional wallpaper hangers is to lay out a piece of wallpaper on a worktable (glue-side facing up), place the paint roller brush inside a large water-filled bucket, and then roll the brush over the paper until it's soaked.

No matter which method you choose, once you saturate the paper, it will quickly turn to glue, so you

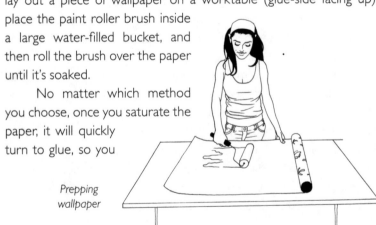

Prepping wallpaper

need to work quickly. Fold the paper in half (glue side to glue side), which is called "booking," with the ends meeting. Find a place where you can allow the paper to rest "booked" for 3 to 5 minutes. Booking wallpaper is really important because it allows the paper to rest so that it can expand. Otherwise, bubbles will form.

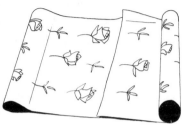

Booking wallpaper

Hanging the Wallpaper

Stand on a ladder or stepladder and, starting at the top of the wall, align the paper with the pencil or chalk line running the length of the wall and press the paper with both hands. It may take several attempts to perfectly align the paper, so don't worry if you don't get it right the first time—the paper won't dry immediately.

Aligning wallpaper

Use the wallpaper hanger to brush the entire piece of wallpaper, from side to side, up and down, and at angles. Wipe the wallpaper seams with the dampened sponge, and then run the roller up and down the seam.

This is where your helpful friend can be *really* helpful!

If a seam or corner isn't sticking, dampen the sponge and wet the area. Run the roller over it. If this still doesn't work, then put a dab of seam adhesive on your finger, gently lift up the edge of paper and apply the glue. Run the roller over the area again.

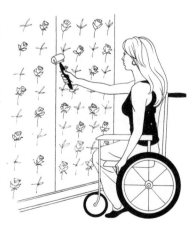

Rolling seams

W et wallpaper can tear very easily, so handle with care.

*I*t's helpful to clean the sink, faucets, etc., as you complete each wall so that any splattered adhesive doesn't stick.

If there are bubbles, wipe the area with the wet sponge and then use the wallpaper hanger to smooth out the section.

Cut away any excess paper at the top and bottom of the walls using a razor blade with the metal spackling knife acting as an edger. Each blade will be good for only 1 or 2 cuts, which is why we told you to buy in bulk. Be sure to have a garbage can nearby to throw the used blades into.

If you wallpapered a bathroom, let the wallpaper dry at least 24 hours before showering. And wait a day or two before hanging heavy mirrors and pictures.

Cutting away excess wallpaper

Tools Needed

Tape measure

Wallpaper

Wall covering primer

Drop cloths

6-foot worktable

Long level or chalk line

Ladder with shelf

Pencil

Scissors

32-inch wallpaper tray

Paint roller frame with rollers

Deep bucket

12-inch hanger and smoother

Large sponge

Wallpaper roller

Smoother (plastic tool)

Helpful friend

Seam and repair adhesive

Razor blades

Cleaning products

Metal spackling knife

Removing Wallpaper

When Sally bought her first house, she fell in love with the home's curtains, chandelier, and Sub-Zero fridge. Who knew that the only thing that would convey was the 1960s groovy wallpaper?

The uglier the wallpaper, the harder it is to remove—even if the walls were sized. Ugly wallpaper just sticks better. It's a fact.

You may already be familiar with the most common method of wallpaper removal, which requires a scorer and a steam machine. Well, we have a different technique that will save you time, will keep your walls from being damaged, and is reusable! It's called *Wallwik*.

Buying Wallwik

At the time of this writing, Wallwik is available only online (www.wallwik.com).

Before ordering the appropriate product, you'll need to know what type of wallpaper you'll be removing (vinyl, grass cloth, foil, or paper), whether it's strippable or peelable vs. all other types, and the size of the room.

We recommend that you order extra sheets of Wallwik because you'll get the job done faster with more.

Getting Started

The night before you plan to remove the wallpaper, take everything off the walls and countertops.

Set up a ladder or stepladder near the wall, using the appropriate safety measures.

This is a project that you can definitely do yourself, but if you can enlist a helpful friend you will drastically cut the time of the project.

Use the scorer to lightly score the walls, being careful not to perforate the drywall. If you have a wallpaper border, you'll need to score a little harder to perforate the wallpaper underneath it.

If the walls have wainscoting on the bottom half with wallpaper above, you'll need to protect the wainscoting. Cover the wainscoting with a plastic drop cloth, or large garbage bags that are cut, and tape them to the trim.

Place the drop cloth on the floor and lay old, rolled towels around the perimeter to catch any drippings.

Scoring the walls

It's important to prevent water with solution from entering light switches and receptacles to avoid electrical shocks and circuit shortages. Therefore, we recommend that you cover the light switches and receptacles with masking tape.

Removing the Wallpaper

Fill the gallon container with warm water and empty it into the bucket along with 1 to 2 ounces of the power solution. Dip the sprayer into the solution and allow it to fill up.

Drop the sheets into the bucket and allow them to soak for a few minutes. Remove a sheet and wring it out so that it's not dripping but still wet. The biodegradable solution is as mild as shampoo, so you can do this without wearing rubber gloves.

Starting at the wall with the most uninterrupted space, hang the sheet vertically from the top down. Where necessary, fold or cut a sheet with scissors to fit the area

Note: The sheet is easier to cut when it's wet.

Working from the top down, smooth out any bubbles and wrinkles. Repeat with the other sheets. If the sheets fall off, use some small tacks to keep them up.

*Applying
solution-soaked
sheets to wall*

*Peeling off
wallpaper*

Note the time on your watch. During the next hour, spray the top third of the sheets every 20 minutes (3 times per hour).

Remove the Wallwik sheets and put them back into the bucket. Starting at the top or bottom of a sheet of wallpaper, lift it up and off . . . hopefully in one sheet! Throw the wallpaper into the garbage bag.

If you've left some wallpaper and/or wallpaper backing behind in the process, place a wet Wallwik sheet back on it, spray again, and pull it off in 30 minutes.

If there are small stragglers, use a wet sponge to rub them off or a plastic putty knife to scrape them off.

Once a wall has been completely cleared of wallpaper, use a wet sponge to remove any residue. Wipe down the walls with a clean, dry towel, if necessary.

If you're working in a large room, we recommend that you use a fresh batch of water and solution halfway through the project.

When the job is done, allow the sheets to dry completely before storing. The sheets are reusable, so save them for another room or share them with a friend.

Tools Needed

Wallwik kit (includes scoring tool, power solution, Wallwik sheets, pressure sprayer, skimmer, and instructions), plus extra sheets

Stepladder or ladder

Plastic drop cloth (if you have wainscoting)

Large garbage bags

Masking tape

Drop cloth

Old towels

Helpful friend (to cut the time)

Empty gallon container

Bucket

Scissors

Small tacks

Large sponge

Plastic putty knife

Painting Interior Walls and Ceilings

For some of you, the following information will be a *primer* on painting. For others, it will let you *brush* up on your techniques. No matter what, you'll see that there's a lot more to painting than meets the surface.

Did you know that there are instructions on paint cans? And you thought you could just slap some paint on a wall and call it a day, right? Painting is a relatively inexpensive way to bring a dull room back to life—that is, if you do it yourself. If you've ever gotten a quote to have some rooms painted, you probably wondered, how can a contractor charge that much? Well, the answer is that the cost includes the hours of necessary prep work—clearing the rooms, patching drywall, cleaning the walls, applying painter's tape—and that's all before opening a can!

There are a lot of tools and necessary steps—yes, we said *necessary*—to ensure that the finished product will look as if you paid a professional painter to do it. Just be sure to schedule enough time for prepping *and* painting.

Note: If your home was painted before 1978, you may have lead paint on the walls. Before beginning this project and possibly risking your health, visit www.hud.gov to learn the appropriate steps to take.

Buying Paint

Before you purchase paint, you need to know two things: (1) the size of the room, and (2) what you want the paint to do for you. Knowing the dimensions of the room will enable you to buy just the right amount so that you won't have to run to the store midway through the job, or get stuck storing the extra gallons. Paint is *not* returnable—used or unused—no matter where you buy it.

Finding the Room's Square Footage

A gallon of paint will typically cover 350 to 400 square feet, but be sure to check the label on the paint can because manufacturers differ with their estimations. To find a room's square footage, measure the length and width of the room using the measuring tape and multiply the two dimensions. For example, if the length of the room is 12 feet and the width is 8 feet, then the square footage is 96 square feet. (This would be the same measurement for the room's ceiling.)

To get a more accurate estimation, including number of windows, height of walls, and even width of molding, visit www.lowes.com to use its paint calculator.

Choosing the Right Paint

Different paints serve different purposes, and we're not talking about color. We're talking about finish. For example, if the room gets a lot of traffic, you'll want the finish to be either *satin* or *eggshell* for easy cleanup. If the walls are showing their age, then use a *flat* finish, which is more forgiving of imperfections. For trim work, wainscoting, and doors, the best finish is *semigloss*. And for walls that are newly constructed and have never been painted, or walls that have been repaired or have a stain or a semigloss finish that needs to be covered, *primer* is required. Primer also works as an adherent for the topcoat so that it lasts longer.

Prepping the Room

Prepping a room can be very time consuming, but probably not for the reason you think. It's not just the size of the room that eats up time but also the amount of items in the room and the condition of the walls.

Note: If drywall repair is involved, you'll need to add at least another day to the project timetable for patching the drywall and painting primer on the walls and/or ceiling.

When choosing paint for a bathroom, look for one with a mildew-resistant formula.

Remove everything from the walls and closets and move as much of the furniture out of the room as possible. All remaining pieces should be pushed to the center. Dust the baseboards and ceiling fans, and then vacuum the flooring.

Cover the furnishings with the plastic sheets or old bedsheets. Place drop cloths (not plastic) on the flooring so that it's completely covered. Now spread newspapers on top of the cloths around the perimeter of the room.

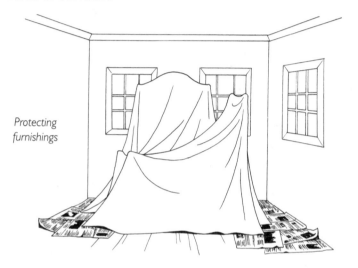

Protecting furnishings

Now that the walls are bare you'll have a better sense of what work lies ahead for you. Using the claw of the hammer, remove all of the hanging nails. Walk around the room with a roll of painter's tape and attach a piece next to any holes, cracks, or dents that will need to be fixed. Don't forget to check the ceiling and inside the closet, too. Repair any damaged drywall (see our first *Dare to Repair* book) and be fastidious about cleaning up afterward, because any dust that lingers will end up on your freshly painted walls.

Speaking of cleaning up, now that the walls have been repaired, you need to clean them. Clean them, you ask? Isn't paint going to make the walls look clean? The reason for cleaning the walls is to

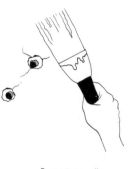

Prepping walls

remove the dirt so that you're not painting over it, and also because a clean wall will allow the paint to adhere better. And if you have pets in your home, this is really important to do because otherwise you'll find yourself painting over cat and dog hairs that landed on the walls.

Following the manufacturer's instructions, add TSP-PF detergent to the warm water in the bucket. Wearing the disposable gloves, dip the sponge mop into the bucket and wipe down the walls (and ceiling, if you're painting it), beginning at the top and working your way to the bottom. Dip the rag into the bucket and clean the windowsills, baseboards, doors, and trim, too. Once the room has been cleaned, dump out the water in the bucket, fill it with clean water, and rinse the sponge mop thoroughly. Now wipe down everything with the clean water. Let dry.

Using a flathead screwdriver, remove all of the electrical covers, including switch, receptacle, cable TV outlet, and phone jack, and place them and their fasteners into a large plastic baggie. Apply painter's tape over the switches, receptacles, etc.—do not paint them! Not only will they look bad, but painting them may lead to you getting an electrical shock.

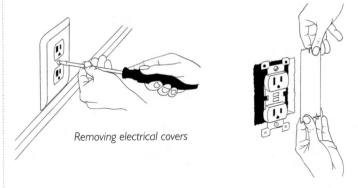

Removing electrical covers

Applying painter's tape
to receptacles

If you're painting the ceiling, remove any ceiling medallions (you may need to remove the light fixture) and tape around any light fixtures or ceiling fans.

Painter's tape works great on trim, doors, and ceiling fixtures. However, we don't recommend using it on walls or ceilings because we find it very time-consuming and messy, and it can peel off the old paint. Instead, we suggest using a paint tool called an edger and using a paint technique called "feathering" with a paint brush.

Applying painter's tape to trim

Tools Needed

Old clothing and hat

Dust mask

Tape measure

Dust cloth

Vacuum cleaner

Plastic painter cloths or old bed linens

Drop cloths

Newspapers

Tack or claw hammer

Painter's tape

TSP-PF detergent (phosphate free)

Bucket with warm water

Disposable gloves

Clean sponge mop (on an extension pole, if necessary)

Clean rag

Flathead screwdriver

Plastic baggies (large and small)

Edger

Extension pole

Paintbrush

Painting the Room

Loading the Brush and Roller

Using the paint can opener or flathead screwdriver, pry open the lid and remove it.

Note: If you'll be using 2 or more gallons of paint for the room, pour each gallon into a clean 5-gallon bucket and mix the paint with a paint stick. This will ensure consistency in paint color. Pour any remaining paint into the empty gallon containers and secure properly.

Insert the liner into the paint tray and pour the paint into the well.

Place the lid back onto the can and gently tap it closed with the hammer. Wipe off any excess paint from the can. Dip the brush, edger, or roller into the paint, and wipe any excess on the ridge part of the tray. Too much paint will cause drips on the wall and floor and spatter marks on you.

Pouring paint into liner well

Getting Started

There is a definite order to painting a room: (1) ceiling; (2) primer, if necessary; (3) walls; (4) windows and doors; and (5) trim.

Ceiling

One of the biggest mistakes homeowners make is that they forget (or refuse) to paint the ceiling. Who can blame them, especially if it's vaulted or cathedral?

Painting a ceiling typically requires a ladder, which can be challenging. But if you don't do it, the dirty ceiling will make the freshly painted room look dismal. So, get over your aversion and just think of every room as having five walls—one of which is above your head.

When choosing paint for the ceiling, look for one that is tinted. This special paint goes on purple or blue so that you can see where you've missed, but it will dry white.

Before you begin to paint, attach the edger to one of the extension poles, and the roller to the other extension pole. Begin painting the ceiling in a corner with the edger. Be careful not to do too big an area or the paint will dry and you'll see the lines of the edger. Instead, work in sections, using the edger and then the roller. Paint the rest of the ceiling using the roller, making either the letter *M* or *W*, and then filling in. You'll probably need only one coat for the ceiling—yippee!

*I*nvest in a high-quality ladder and stepladder for your personal safety.

Using a paint edger   *Applying paint*

Primer

Be aware that primer is thicker than regular paint, and therefore you'll probably need more of it to get the job done. Primer can be used on the walls and ceiling, but is not necessary for trim and doors. Follow the same rules as for painting.

Walls

Painting walls requires a method called "cutting in" or "feathering." This is when you take a 2-inch *straight* brush and make small sweeping motions from the corners and edges out, about a few inches. Cut in only about 3 feet at a time, which will allow you to paint over the area with a roller brush and not see any lines.

Cutting in

Another great tip is to use an edger for painting around windows and doors. Again, make sure to immediately paint over the area with a roller.

Fill in the rest of the wall area with a roller, using the M or W. Remove the tape while the paint is wet.

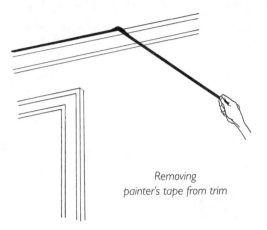

*Removing
painter's tape from trim*

Let the walls dry about 3 hours and then decide if you need another coat.

Windows and Doors

When painting windows made of wood, use the sash brush, which is angled, and follow the grain of the wood. With the windows open, start at the top of the sashes and work your way down to the sill. Then paint the window casing, again starting at the top. And then paint the sill.

To paint a flat door, use a roller. To paint a door with raised panels, use a brush to paint the panels first, and then the entire door. Don't forget to paint both sides and the edges.

Trim

For painting trim and baseboards, use a sash brush. Make sure not to overload the brush and paint following the grain of the wood. Use a paint shield or a piece of cardboard to push under the baseboard so you can paint all of it.

TRICKS OF THE TRADE

- **Keep the room well ventilated.**
- **Use a disposable liner for the paint tray.**
- **Pour the paint only into the well of the tray so that the ridged part can be used to remove excess paint.**
- **Use a synthetic brush for latex paint and natural fiber bristles for oil paint.**
- **Don't overload the roller or brush.**
- **Use an angled brush for tight corners and small areas.**
- **If you take a break from painting, wrap the roller and brushes in plastic wrap to prevent them from hardening.**
- **Always put the lid back on the can, even if the job is not done.**
- **Ask for a paint can opener when purchasing paint. They're typically free.**

Cleanup

It's important not to remove any of the painting tools until you take a final look around the room, because it's so much easier to touch up an area with brushes and rollers that are still wet and the paint can within reach. Otherwise, you may just want to overlook the flaw and throw a picture over it!

Rollers and edger pads can be thrown away, but if you purchased quality brushes, then you should clean them. For latex paint, rinse the brushes with warm soapy water, squeeze out the excess water, and hang to dry. For oil paint, use a coffee can or glass jar and pour enough paint thinner into it to cover the bristles. Soak the brushes, and then rinse with warm soapy water. Run the paint comb through the bristles and hang the brush to dry.

Tools Needed

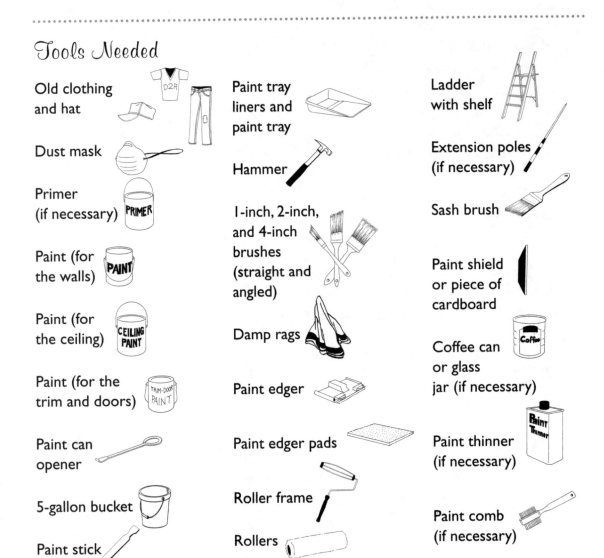

Old clothing and hat

Dust mask

Primer (if necessary)

Paint (for the walls)

Paint (for the ceiling)

Paint (for the trim and doors)

Paint can opener

5-gallon bucket

Paint stick

Paint tray liners and paint tray

Hammer

1-inch, 2-inch, and 4-inch brushes (straight and angled)

Damp rags

Paint edger

Paint edger pads

Roller frame

Rollers

Ladder with shelf

Extension poles (if necessary)

Sash brush

Paint shield or piece of cardboard

Coffee can or glass jar (if necessary)

Paint thinner (if necessary)

Paint comb (if necessary)

Repairing a Large Hole in a Wall or Ceiling

When Sarah was moving into her house, she was shocked to see that the original homeowners had not patched a large hole, as promised. Sarah could only hope that the door *did* hit them on the way out.

First, you pretended it didn't exist. Then you tried hanging a picture over it, but it was an obvious cover-up. And then you got a few quotes to have it fixed, which almost made you put a *new* hole in the wall. We promise you that the hole is much harder to look at, day in and day out, than actually repairing it!

Little dents can be filled with a dab of Spackle, and small holes can be fixed with a drywall patch, but a hole that's larger than 6 inches needs to be replaced with a new piece of drywall. Why? Because drywall, *as you now know*, isn't very resilient, so putting a patch over a large hole would be like putting cardboard on a broken window—it won't last long. Therefore, you need to replace the damaged area with a new piece of drywall.

Buying Supplies

When you're buying Spackle (a.k.a. joint compound), buy a small container, not a large bucket, and read the label carefully. Avoid any product that states that it dries in 20 minutes, because you won't be able to sand it.

And if you're panicking about how you're going to schlep a big sheet of drywall home from the hardware store, don't. You can actually purchase a *piece* of drywall, typically 2' × 2', instead of the regular size, which is 8' × 4'. Just be sure to have something like some old towels in your car to set the drywall on because it can be as chalky as its bigger version.

Getting Started

There are a few things you need to know prior to starting this project. One is that if you're repairing a hole in a ceiling, you'll need to use a ladder, wear a tool belt, follow the rules for ladder safety (see our first *Dare to Repair* book), and have a helpful friend assist you while you're up there.

This project creates a lot of dust, so be sure to wear a dust mask and safety glasses, and clean up immediately.

Also, it's very important that the drill be *fully* charged, because if it's only partially charged, it won't have enough power to insert the drywall screws into the studs, and you *don't* want to try doing that by hand. So, charge it up the night before.

Removing the Drywall

Place the drop cloth on the floor directly under the hole.

Using the flashlight, take a look inside the hole for any electrical wires. If there are wires, then be extra cautious when cutting the drywall. Now locate the 2 studs, one on each side of the hole (you may need to use a stud finder). The replacement piece of drywall will need to be attached to these studs.

Use the pencil to mark on the drywall above the center of each stud. Then, using a carpenter's square or the long level, draw a line from the center of one stud to the center of the other. Now draw a line on each end so that you have a rectangle. Once you're done, measure the length and width of the rectangle, and write down the dimensions.

Measuring damaged drywall

Use the drywall saw to cut out the rectangle, being careful to stay on the line. Cutting drywall isn't difficult, but it can be tedious and it's definitely dusty. If you're sawing into the ceiling, have your helpful friend hold the box underneath the hole to catch as much of the dust as possible.

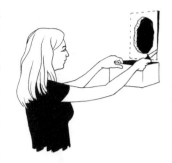

Using the drywall saw

When you reach a stud, you'll need to switch out the drywall saw for the utility knife to make it easier to remove the drywall from the stud.

Once you've completely cut around the rectangle, remove it and throw it out. Clean up the area.

Studs, which are the boards that run vertically from the floor to the ceiling throughout your house, are typically spaced 16 inches apart on center. On center means that the measurement is taken from the center of the boards, not the edge. A joist is a wood beam that supports the ceiling, and it's typically 16 inches apart, on center, from the next joist.

Installing the New Drywall

Place the new piece of drywall on the drop cloth. With the measurements that you wrote down, use the carpenter's square or the long level to draw a rectangle. It's best to incorporate the edges of the drywall in the drawing, especially the finished edge, so that you have only 2 lines to cut.

Using the utility knife, score the lines of the rectangle. Now it's time to do it like the pros—you're actually going to bend the drywall at the long line. Once it's bent, use the utility knife to cut through. *Who knew, right?* Repeat with the remaining side.

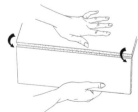

Bending the drywall

Insert the rectangle into the hole, with the paper side of the drywall facing out. It should be a snug fit. If it's too snug, trim the edges of the hole with the sanding block or drywall saw. Once you get the piece fitted properly, you can begin inserting the drywall screws.

Inserting drywall into the hole

Check that the drill is in forward mode, and insert the bit. Line up a drywall screw in a corner so that when you drill it, it will go through the drywall and into the stud. Drill the screw into the stud, going deep enough so that the drywall is secured without breaking the paper. Repeat on the remaining 3 corners.

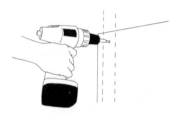

Drilling into the stud

Finishing the Area

The next step is to cover the 4 joints (where the 4 sides of the new drywall piece meet the 4 sides of the existing drywall) with mesh tape. Roll out the length that you need, tear it off, and secure it to the joint by running your hand over it. Repeat for the other joints, being careful not to overlap the tape.

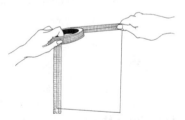

Covering joints with mesh tape

Using the smaller putty knife, apply a generous coat of Spackle to the taped area, spreading it smoothly and feathering it out at the sides. Let the area dry completely, typically 24 hours.

Wearing a dust mask and safety glasses, use the sanding block to sand the spackled area, being careful not to get sanding-happy

Applying spackle

because then you'll have to start from the beginning! Wipe down the area with a cloth and remove the dust on the drop cloth, too.

With the larger putty knife, apply a skim coat of Spackle (a.k.a. finishing coat) over the area, spreading it smoothly and feathering it out at the sides. Wait for the area to completely dry before sanding again, about 24 hours. Wipe the dust off the area and remove the dust on the drop cloth, too.

Now you're ready to paint!

Tools Needed

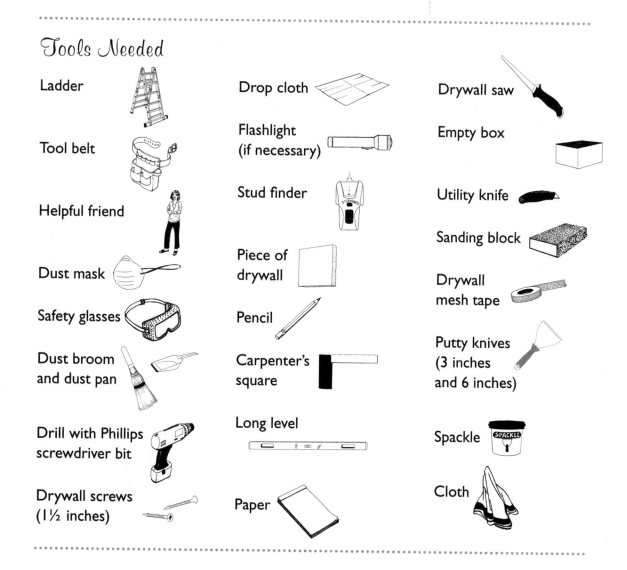

Ladder

Tool belt

Helpful friend

Dust mask

Safety glasses

Dust broom and dust pan

Drill with Phillips screwdriver bit

Drywall screws (1½ inches)

Drop cloth

Flashlight (if necessary)

Stud finder

Piece of drywall

Pencil

Carpenter's square

Long level

Paper

Drywall saw

Empty box

Utility knife

Sanding block

Drywall mesh tape

Putty knives (3 inches and 6 inches)

Spackle

Cloth

Replacing a Ceramic Tile

An area with many cracked tiles could be a sign of an underlying problem ... literally. The number one cause would be that not enough mortar or mastik was used when the tiles were installed.

Even Jessica Fletcher couldn't have figured this one out. Trish came home from work to find a single cracked tile on the kitchen floor. There was no evidence of what caused it. Both parties looked guilty, but Trish couldn't prove whether it was the cat or the dog, and the fish wasn't talking. Trish knew the real mystery was going to be how to replace the *one* tile!

A damaged tile is always in the most noticeable place on the floor or wall. But if you're the type of person who thinks a chip or crack adds character, then we say, *live with it*. Here's why.

It may seem like a blessing at first that only one tile was damaged, but that's only if you have a replacement tile. If not, depending on the age, color, and style of the tile, you may have a hard time finding an exact match. You also have to purchase grout that will complement the existing grout. And don't even think about hiring out for this job, because no one will come out to replace *one* tile.

But ... if you're the type of person who will see the crack even in your dreams, then have we got a project for you!

Buying a Replacement Tile

In most situations, when you need to get an exact match of something, you bring a sample to a store—but not with tile. When it comes to a broken tile, you shouldn't remove it until you know for sure that you can replace it.

What you can do is measure the existing tile, note the color and texture, and visit local tile and home improvement stores to purchase a variety of single tiles that may be a match—they can always be returned with a receipt.

If you did find a match, great! Be sure to buy a boxful for possible future use.

If you can't find an exact replica, then you may want to consider doing nothing, or think about creating a new design on the wall or floor, or replacing the entire thing.

Buying Tile Spacers

If you're replacing a tile on a wall, you'll need to purchase tile spacers.

Tile spacers are little white x's that are designed to fit into the mastik, in between the tiles, to create the exact space desired for grout. So, if you want the grout to be ¼ inch wide, which is the largest width for grout, look on the tile spacer bag for the words "¼ inch." Tile spacers come in ¼ inch, ⅛ inch, ¹⁄₁₆ inch, and ¹⁄₃₂ inch, and in bags of 200. It's really important to get the correct size tile spacers or you'll throw off the entire design of the backsplash.

Measure the width of an existing grout line to get the appropriate size.

Buying Mastik or Mortar

The type of adhesive you need for the tile—either mastik or mortar—will be determined by where the tile will be installed. For example, if the tile will be on the floor, then you need mortar (a.k.a thin-set), which is cement-based. If the tile is on a wall in a kitchen, then you need mastik. But if the tile will be on a shower wall, then you'll need mortar. Always read the label on the product to determine which type of adhesive you'll need and the amount of area it will cover.

You don't have to worry about picking the right color because once the tile is installed, you'll never see it.

Mastik comes premixed and ready to use, but because mortar is cement-based, it comes as a powder in a bag and needs to be mixed with water. Since you're replacing only one tile, look for the smallest size bag.

Tiles are pretty indestructible. In fact, if tiles are installed properly, they're more likely to chip rather than crack.

Buying Grout

Grout is a bonding material that fills in the gaps between tiles. It helps to keep the tiles in place while also keeping moisture from getting under the tiles.

Grout comes sanded and unsanded in both a premixed formula that's ready to use and in a cement-based formula that needs to be combined with water.

You have to read the labels on the different types of grout to make sure that you're getting the right one for the job. If you're not sure, consult a store clerk, and if you're still not sure, then call the grout manufacturer's 800 number right there in the store (the number is always listed on the container).

Regarding color, the rule of thumb is to have the grout match the old grout, but it's going to be tough to match it perfectly because the existing grout has aged and probably discolored a bit. One option is to buy two shades (from the same manufacturer) that are similar to the existing shade and make samples until you find the right combination. Just be sure to let the samples dry completely to know which one will work best.

Buying a Rotary Tool

We found that the best tool to use for removing grout is the Dremel cordless rotary tool, with a grout removal bit. We figure that if you have to use a tool, why not use one that can get the job done quickly while you're having fun using it? Plus, this tool comes with lots of cool attachments, along with a book of ideas for using it around the home and its own little carrying bag. Be sure to read the manufacturer's directions for using the tool and attachments.

The only downside to using this tool is that it sounds eerily similar to when you're sitting in the dentist's chair. The upside is that you're on the other end of the drill this time!

Getting Started

Read the manufacturer's directions for using the rotary tool and attachment and follow the guidelines for charging the battery.

Removing the Grout

The object of removing a cracked tile is to avoid damaging adjacent ones. So, the Ceramic Tile Education Foundation recommends that you remove the grout *before* removing the tile. By doing it this way you'll avoid putting pressure on the surrounding tiles, which can cause them to crack.

Wearing the safety glasses and dust mask, turn on the drill to the low setting—this will allow you more control than using the high setting—and with the tip of the drill bit on the grout, move it back and forth on one line of grout until you see the grout line

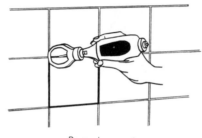

Removing grout

turning dark. A dark line means that you've removed the grout (the grout gets pulverized into dust). Continually wipe away the dust with a wet cloth so you can see your progress. Repeat on the other 3 lines, being careful not to veer onto a neighboring tile.

Removing the Broken Tile

As we mentioned above, the goal of removing a tile is to not damage other tiles in the process. So, be careful during the next step.

Place a rag or cloth over the damaged tile and use the hammer to smash it into pieces. You may have to

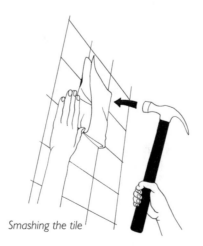

Smashing the tile

smash it more than once, so, *have at it, sister!* Wearing work gloves, remove all the pieces and discard.

Removing the Old Mastik or Mortar

Before installing the new tile, you need to remove all the old mastik or mortar so that the new tile will adhere properly.

Wearing the safety glasses and dust mask, chip away the adhesive, using a hammer and chisel. Remove the pieces and vacuum the area.

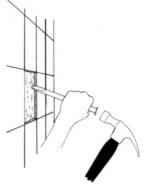

Chipping away the adhesive

Installing the New Tile

If you're using mortar as the adhesive, you can just "back butter" the tile, kind of like buttering a slice of bread—you do it on only 1 side. We actually found it easiest to just use a butter knife to spread the mortar onto the back of the tile. Just be sure to clean the knife *before* putting it into the dishwasher!

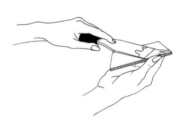

Back buttering a tile

If you're using mastik, you *can't* back butter the tile because mastik dries by evaporation. Therefore, use a small putty knife to scoop a bit of mastik into the opening and then use the ridged edge of the small trowel to prepare the area (the trowel will give you just the right amount of mastik).

Push the tile into the opening, move it side to side a bit, and gently tap it with the rubber mallet. If the tile is on a wall, place the appropriate size tile spacers in between it and the

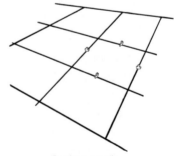

Replacing a tile

surrounding tiles to keep it in place. Wipe away any excess with a damp cloth. Wipe off the putty knife completely before washing it.

Follow the adhesive manufacturer's guidelines for drying.

Applying Grout

Read the manufacturer's directions for the grout.

If the tile is on a wall, spread a drop cloth close to where you'll be working to protect the area from any grout droppings.

Remove the tile spacers, if applicable, and save for future use.

If you're using premixed grout, the container may not be big enough to fit the rubber float, so you may need to use the plastic putty knife to dig into it. Transfer the grout from the putty knife to the bottom of the rubber float.

Tilting the rubber float at a 45-degree angle, spread the grout over the tile, making sure that it's getting in between the joints around the tile. You may find that having both hands on the rubber float works best. Once the area has been grouted, tilt the rubber float at a 90-degree angle and go over the area again.

Applying grout with a rubber float

Follow the manufacturer's directions for how long to wait before cleaning the tiles (typically 15 to 20 minutes) and how long it will take the grout to set (typically 24 to 48 hours).

Wipe the tile with a damp sponge to remove any excess grout. If there is any hazing on the tile, wipe it with a soft cloth. Follow the grout manufacturer's guidelines for drying.

Tools Needed

Replacement tile

Tile spacers
(if necessary)

Tape measure
or ruler

Mastik or
mortar

Grout

Dremel
cordless
rotary tool

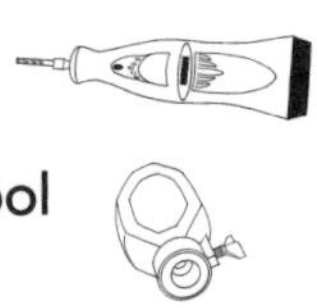

Safety glasses

Dust mask

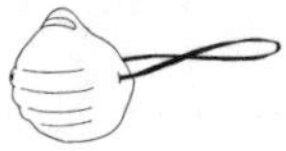

Wet cloth

Hammer

Work gloves

Chisel

Vacuum cleaner

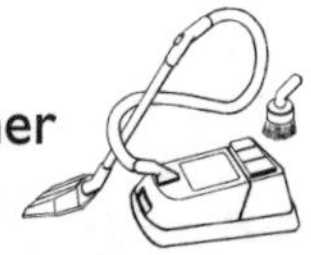

Butter knife

Small plastic
putty knife

Small notched
trowel

Rubber mallet

Drop cloth

Rubber float

Sponge

Soft cloth

Installing a Ceramic Kitchen Backsplash

The kitchen is the heart of the home, and Pam knew that hers needed some **CPR. But the price of having it remodeled just about gave her a heart attack, so she opted to keep the cabinets and install a tile backsplash instead. Pam says that doing the job herself brought her kitchen back to life and was a great shot in the arm for her self-confidence.**

Before you begin this project, answer these questions: (1) Have you ever tiled before? (2) Are you artistic? (3) Are you a professional contractor? If you answered no to all three questions, then this project is for you!

That's right. Installing a ceramic backsplash is for anyone who is a nonartistic novice handywoman. It's really straightforward. And if you answered yes to any of the questions, well then, this project will be even easier.

Now, having said that, we do recommend that you enlist the help of a friend (or two) to make things go faster and to provide encouragement, because just the thought of doing this project can be overwhelming. But we promise you that once you get through the nitpicky stuff and start adhering the tiles, you'll just fly through it. And the excitement of seeing the drab wall turn into a work of art is thrilling. You will be the "Queen of the Tiles."

Designing the Backsplash

Tiles come in a wide variety of shapes, sizes, textures, and colors, so be sure to bring along a paint chip or swatch of wallpaper from the kitchen when you go shopping. We recommend buying a few single tiles so you can bring them home to get a real feel for how the backsplash will look. Just be sure to take into account that small tiles (e.g., mosaics) are difficult to cut, and a design that requires tiles to be installed on the diagonal will entail a lot of cutting. Having said that, it's your kitchen, and you've got to love the tile and the design.

Once you've determined the size of the tiles and design, you need to decide how much grout you want showing in between each tile. It can range from a small amount ($\frac{1}{32}$ inch) to a wide space ($\frac{1}{4}$ inch), with $\frac{3}{16}$ inch being the most common.

Using the tape measure, find the dimensions of the backsplash area. Transfer the measurements of the backsplash (including the grout) and design it *to scale* on a piece of graph paper. We recommend that you make a few copies of the design once you're done.

Buying the Tiles

Count the number of plain tiles and the number of tiles with a motif or special shape. Take this information, along with the measurement of the backsplash, to the store to purchase the tiles. We recommend that you purchase extra tiles (typically 10 percent more) for the mistakes that will happen, as well as to have on hand for any future needs.

Buying Mastik, Grout, Tile Spacers, and Caulk

Mastik

The type of adhesive you need for the tile—either mastik or mortar—is always determined by where the tile will be installed. Because the tile will be installed as a backsplash, we chose mastik as the adhesive.

We like working with the ready-to-use mastik because there's no mess and no fuss. Read the label to know what size area the mastik will cover. When in doubt, buy an extra tub because you may be able to return it with a receipt, if it goes unopened.

Regarding the color, mastik typically comes in a variety of colors, but you really don't have to worry because once the tile is installed you won't see it.

Grout

Grout is a bonding material that fills in the gaps between tiles. It helps to keep the tiles in place, while also keeping moisture from getting under the tiles.

Grout comes sanded and unsanded in both a premixed formula that's ready to use and in a cement-based formula that needs to be combined with water.

You have to read the labels on the different types of grout to make sure that you're getting the right one for the job. If you're not sure, consult a store clerk, and if you're still not sure, then call the grout manufacturer's 800 number right there in the store (the number is always listed on the container).

Regarding color, the rule of thumb is to have the grout match the tile, but that's your choice.

Be sure to check the packaging to find out the area that the grout will cover.

Tile Spacers

Tile spacers are little white x's that are designed to fit into the mastik, in between the tiles, to create the exact space desired for grout. So, if you want the grout to be ¼ inch wide, which is the largest width for grout, look on the tile spacer bag for the words "¼ inch." The tile spacers come in ¼ inch, ⅛ inch, 1/16 inch, and 1/32 inch, and in bags of 200. It's really important to get the correct size tile spacers or you'll throw off the entire design of the backsplash.

Caulk

Because you're installing a kitchen backsplash, you'll need to choose a latex-based (water resistant) caulk and match the color of the grout as closely as possible. Caulking is done after the grout has dried.

Tools

You will absolutely, positively, need a wet saw. Don't even think about using a tile cutter or tile nipper. Trust us. Either invest in a wet saw (less than $100) or rent one for the day.

Some other tools that you'll need and probably don't have are (1) a trowel and a notched trowel, and (2) two sawhorses and a piece of plywood that's large enough to support the wet saw.

Prepping the Walls

The wall needs to be in really good shape so that the mastik will adhere properly. Therefore, if the wall needs to have wallpaper removed (see page 113) or holes patched (see page 127), do this in advance.

Turning Off the Electricity

If the backsplash area has electrical receptacles and/or switches, turn off the electricity to that wall at the main service panel. With a flat-head screwdriver, remove the screws to the covers of the outlets and switches and place them all in a plastic baggie. Touch the wires and terminals (the screws on the side of a switch) with the voltage tester. If it emits a beep, then the power was *not* turned off. Keep trying until no beeps are emitted.

Use the screwdriver to loosen the screws that hold the receptacles/switches to the electrical box. Pull the receptacles/switches out just enough so that they will be flush with the new tile surface. You can reuse the covers, you'll just need longer screws.

Getting Started Outside

It's best to cut the tiles outdoors. So, lay a plastic drop cloth on the deck or patio and set up the sawhorses with the piece of plywood on top. Place the wet saw on the plywood and attach an extension cord, if necessary. Fill the tray of the wet saw with water.

If you need to do any diagonal (half) cuts, do these before you get started inside. Use a black marker and a ruler or a carpenter's square to create a diagonal line on the back of each tile. Wearing the safety glasses, dust mask, work gloves, and work apron, turn on the machine and gently guide the tile (back side up) through the blade.

If you notice that the ends seem to be consistently breaking off, then try cutting a tile about ½ inch down, flip it around, and begin cutting into the other end, guiding it all the way through the cut corner.

Preparing the wet saw

They don't call it a wet saw for nothing. Be sure to wear a work apron to keep dry.

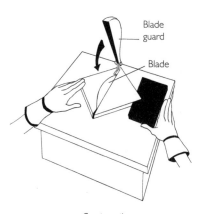

Blade guard

Blade

Cutting tile

Getting Started Inside

Protect the countertop and any appliances that are in the immediate area by covering them with a plastic drop cloth.

You need to begin tiling at the center of the backsplash, along the bottom. To find the center point, measure the length of the area, divide by 2, and mark the spot with a pencil.

Now you need to create a starting line at the bottom of the wall. Place the laser level on the wall and turn it on. Adjust the laser so that the line is where you want the bottom row of tiles

Measuring centerpoint of backsplash

to be. Have your helpful friend use the pencil and the long level to draw on the laser line. This is now your starting line.

Applying the Mastik

Read the manufacturer's directions for the mastik.

Use a flathead screwdriver to open the container of ready-to-use mastik. There's no need to stir the contents. Dip the small trowel into the bucket and spread the mastik onto the starting area. Now use the notched trowel to go over the area, creating a wavy pattern. Try to keep the starting line visible through the mastik.

Apply the mastik only to an area that you know you will cover with tiles within 15 minutes. Remove any excess mastik with the trowel or sponge.

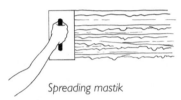

Spreading mastik

Applying Tiles

Starting at the center point on the starting line, place a tile onto the mastik, wiggling it side to side a bit. Knock the tile a few times with the rubber mallet. Place a tile next to the first one, using the same method, and then insert 1 or 2 tile spacers between the tiles. If the tile spacers keep falling out, then you need to adjust the tiles so that spacers remain in place. Remember, these tile spacers are designed to keep the tiles apart the exact same distance so that the grout will look uniform.

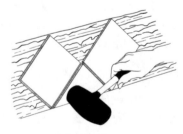

Adhering tile

Once the first row is complete, check it with a level to make sure it's straight. Move on to the next row, starting at the center and working your way to each end. Again, check that the row is level before moving to the next row.

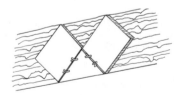

Inserting tile spacers

Wipe off any excess mastik that's on top of the tiles with a wet paper towel. Use a butter knife to remove any mastik that's between the tiles—you'll need this space for the grout.

Making a Special Cut

It's almost impossible to avoid making any difficult cuts—that's why you bought the extra tiles! But don't fret, because we have a great way to make it easy to do.

Take a piece of cardboard (sometimes pieces come with the tiles) and place it against the area that requires a special cut and trace the pattern. Use the scissors to cut out the shape, and trace it with a black marker onto the back of a tile. Now use the wet saw to make the cuts, following the pattern.

Cleaning Up

If you have any excess mastik on the tiles and wall, wipe it off with a wet paper towel. Use a clean putty knife to remove any stubborn areas, if necessary.

To clean the wet saw, unplug the cord from the electrical outlet. Then wipe down the wet saw with wet paper towels. Remove the water tray, which you'll see is filled with a thick paste made from the ceramic shavings. Use the putty knife to scrape out the residue into a garbage can outside and rinse the water tray with a garden hose. Dry the tray with paper towels before replacing it into the wet saw. And dry the wet saw with paper towels before placing it into its box.

Read the manufacturer's instructions before applying the grout. Typically, the wait is 24 to 48 hours after laying the tiles.

Applying Grout

Read the manufacturer's directions for the grout. Spread a drop cloth over the countertop, stove, sink, etc., to protect them from any grout droppings.

Remove the tile spacers and save for future use. Set up 2 buckets of clean water on the counter, along with a large sponge or soft cloth.

Depending on the size of the premixed grout container, you may need to use a small putty knife instead of the rubber float (too big) to dig into it. Transfer the grout from the putty knife to the bottom of the rubber float.

Tilting the rubber float at a 45-degree angle, spread the grout over the tiles, making sure that the grout is getting in between each tile. You may find that having both hands on the rubber float works best. Once the area has been grouted, tilt the rubber float at a 90-degree angle and go over the area again.

Follow the manufacturer's directions for how long to wait before cleaning the tiles (typically 15 to 20 minutes) and how long it will take the grout to set (typically 24 to 48 hours).

Last step!!!

Applying grout with a rubber float

Applying Caulk

Now that the correct time has passed for the grout to set, you can finish the job by applying caulk to the top and bottom of the backsplash.

Read the manufacturer's directions for applying the caulk.

Insert the caulk tube into the caulking gun and snip the end of the tube at an angle. Be careful not to snip too much off because then you'll have a large amount of caulk dispensing from it.

Position the tip at one of the ends of the top of the backsplash, squeeze the handle, and run a bead of caulk to the other end. Wipe away any excess with a fingertip. Repeat on the bottom of the backsplash.

Follow the manufacturer's instructions for drying time.

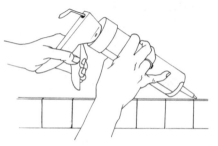

Applying caulking

Tools Needed

Tape measure

Graph paper

Pencil

Tiles

Mastik (ready-to-use)

Grout

Tile spacers

Latex-based caulk and caulking gun

Wet saw

Extension cord

Trowel

Notched trowels of different sizes

Sawhorses and plywood

Flathead screwdriver

Plastic bag

Voltage tester

Long screws (to accommodate extended outlet receptacles)

Plastic drop cloths

Black marker

Carpenter's square or ruler

Safety glasses

Dust mask

Work apron (full size)

Work gloves

Laser level

Helpful friend(s)

Rubber mallet

Paper towels

Butter knife

Cardboard

Scissors

Putty knives

Garden hose

Buckets (2)

Sponge

Rubber float

Storage

There's no such thing as an empty closet.

You moved from a one-bedroom apartment into a three-bedroom house, and somehow you managed to fill every closet. And if you added more space for storage, you'd fill that, too. It's one of those unexplainable phenomena.

We've designed this section not as an organizational how-to but rather as a how-to-get-more from the storage areas you already have and to find storage you didn't know existed.

So, let your inner pack rat out!

Replacing a Recessed Medicine Cabinet

Want to know a great prescription for sprucing up a tired bathroom? Replace the old medicine cabinet. It may be just what the doctor ordered.

Remember that even though a medicine cabinet isn't heavy, you need to take caution in working with it because of the mirror.

Replacing the medicine cabinet is inexpensive and so easy to do that you'll question why you didn't do it sooner.

If you're wondering what to do with all of those outdated medicines, do NOT flush them down the toilet. Ask your pharmacy if it will dispose of them and, if not, follow the EPA's recommended guidelines for doing it yourself.

Buying a Medicine Cabinet

Be aware, recessed medicine cabinets are not one size fits all; in fact, they come in many different sizes. Therefore, you'll need to measure the old one *before* buying a replacement.

The best way to take a measurement is from *inside* the cabinet. First, remove all of the medicines and other paraphernalia. Use the tape measure to find the depth, width, and height of the interior.

Taking measurements

Getting Started

Remove the *new* cabinet from its box and check that it contains the appropriate hardware. Read the manufacturer's instructions before beginning.

Before removing the *old* cabinet, wipe off the exterior, where dust probably collected.

Installing the Medicine Cabinet

With the *old* cabinet completely empty, use a Phillips screwdriver to remove the mounting screws (typically located on the interior sides). Take out the *old* cabinet and place it inside the *new* cabinet's box for safety.

Insert the *new* cabinet into the wall. It should be a snug fit. Place the level on the bottom of the cabinet to check that the cabinet is level. If not, maneuver the cabinet a bit until it is. Use the Phillips screwdriver to fasten all of the mounting screws, being

Removing mounting screws

careful not to overtighten to avoid cracking. Insert the shelves.

Don't put anything into the new cabinet until you've checked every medicine bottle for its expiration date. If any medicine has expired, then properly dispose of it.

If the old cabinet is still in good shape, consider donating it to a Habitat ReStore.

Tools Needed

Tape measure

Pencil

Paper

New medicine cabinet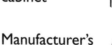

Manufacturer's instructions

Dust rag

Phillips screwdriver

Small level

Installing a Wall-Mounted Medicine Cabinet

*I*f one medicine cabinet just isn't enough to store all the medicines and moisturizers, sunscreen, and wrinkle creams, think about replacing the bathroom mirror with a wall-mounted medicine cabinet. We multitask, so why shouldn't our mirrors?

A wall-mounted medicine cabinet is a great way to increase storage space in your bathroom while doubling as the main mirror in your bathroom. It differs from a recessed medicine cabinet in that it is mounted into the wall studs instead of between them, and it's designed to look more like a mirror than an ordinary medicine cabinet.

This project is a bit more time-consuming than replacing a recessed model, but if storage space is at a premium in your bathroom, it will be well worth your efforts.

Buying a Wall-Mounted Cabinet

A wall-mounted cabinet is more expensive than a regular medicine cabinet, but think of it as getting two things—a mirror and a cabinet—for the price of one.

Use a tape measure to determine the width of the area you'll be installing the cabinet over. When looking to buy a cabinet, you'll need to refer to that measurement to make sure that you get the right size.

Getting Started

Empty the contents of the box of the new cabinet onto a protected surface. Read the manufacturer's instructions to check that all the parts were included.

Finding the Studs

Place the cabinet against the wall at the desired location. Take a pencil and lightly mark on the wall the bottom of the cabinet's center and end points. Lay the cabinet in a safe place.

With the marks as your guide, use a long level to draw a straight and level line the length of the markings.

*Marking bottom of
medicine cabinet*

Now place the stud finder on the wall and run it along the line to locate the studs. Mark the locations of the studs with a pencil. The studs should be 16 inches apart.

Note: If you can't install the medicine cabinet onto a stud, use metal anchors (not plastic ones supplied by the manufacturer). Remember that once the screw has been

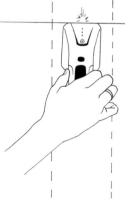

Locating a stud

completely inserted, give it a few additional turns (5 to 7) to fully expand the anchor.

Installing the Mounting Bar

Take the mounting bar and position it along the drawn line. Now place the level on top of the bar to make sure it's level. Using a permanent marker, note the location of the studs directly onto the mounting bar.

Wearing safety glasses, hold the mounting bar over a trash can (this will catch the metal shards) and drill holes at the designated markings.

Don't take the safety glasses off just yet!

Switch out the drill bit (this may be hot) with the bit to create the pilot holes. Drill the required number of holes into the wall.

Position the mounting bar so that the holes are aligned over the holes in the wall studs. Use the Phillips drill bit to drill the screws through the holes in the mounting bar and into the holes in the studs.

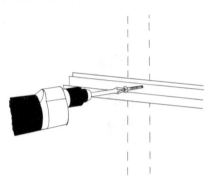

Installing a mounting bar

Installing the Mounting Clips

The provided mounting clips will secure the top corners of the medicine cabinet to the wall once the cabinet has been permanently positioned on the mounting bar.

Temporarily place the bottom of the cabinet onto the mounting bar and use a pencil to mark the location for the top mounting clips.

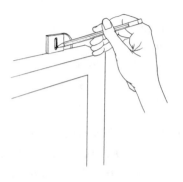

Marking location of mounting clip

> **P**lastic anchors can be unreliable, so use metal ones instead.

Remove the cabinet from the mounting bar and set it safely aside. Wearing safety glasses, drill the pilot holes at the marked locations and remove any debris that falls into the mounting bar.

Loosely install the mounting clips by drilling the screws partway into the holes. (If the holes are not in studs, then you'll need to use anchors.) You'll tighten them a little later when you install the cabinet to the wall.

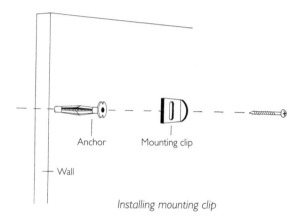

Anchor   Mounting clip

Wall

Installing mounting clip

Installing the Side Mirror Brackets

The side mirror brackets will be used to attach the long, narrow mirrors to the sides of the cabinets.

Peel only 1 side of the 2-sided foam tape and apply it to the bracket. Align the hole on the bracket with the hole on the exterior surface of the cabinet's side. Push the screws into the holes.

Repeat on the other side mirror.

Note: We deviated from the manufacturer's instructions by installing the side brackets prior to hanging the cabinet because side wall clearance prevented easy access. Also, on some models, side-mounting bracket screws are difficult to install and will require some hand strength.

Applying foam tape

Securing Side Mirrors to the Cabinet

Position the bottom frame of the cabinet onto the mounting bar. Slide the top frame of the cabinet into the top mounting clips and tighten the clip screws with the Phillips screwdriver.

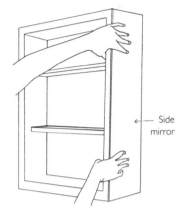

Use rubbing alcohol to clean the back side of the side mirror.

Peel the remaining protective sheet from the double-sided foam tape on the bracket. Carefully align the top and bottom of the side mirror with the top and bottom of the cabinet's side frame.

Apply the mirror with gentle but firm pressure points where the brackets are located.

Repeat the above steps for the other side mirror.

Applying side mirror

Attaching Hinge Screws to the Cabinet

If your new cabinet has adjustable hinges, loosely install (for future up and down adjustments of the cabinet doors) the hinge screws to the inside of the cabinet using the Phillips screwdriver.

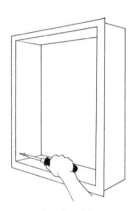

Installing hinge screws to cabinet

Attaching Hinges to the Cabinet Door(s)

Lay the cabinet door on a flat surface. Using a Phillips screwdriver and the provided screws, attach the larger end of the top and bottom hinges to the inside of the cabinet door.

Repeat on the other door(s).

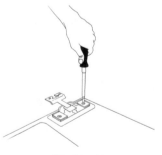

Attaching hinges to cabinet door

Attaching Doors to Cabinet

With your helpful friend holding the door, slip the unattached ends of the door hinges onto the screws installed inside the cabinet. Repeat for the other door(s). Tighten the hinge screws inside the cabinet once the doors are perfectly aligned.

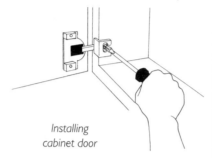

Installing cabinet door

If necessary, adjustments can be made later by simply loosening the hinge screws and sliding the door upward or downward.

Installing Shelves

Now that the doors have been installed, insert the shelf brackets into the desired holes and place the shelves accordingly.

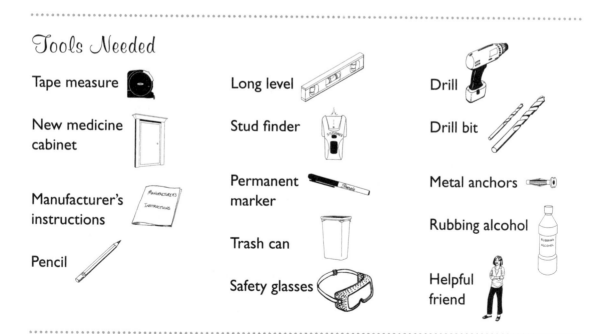

Tools Needed

Tape measure

New medicine cabinet

Manufacturer's instructions

Pencil

Long level

Stud finder

Permanent marker

Trash can

Safety glasses

Drill

Drill bit

Metal anchors

Rubbing alcohol

Helpful friend

Installing a Closet System

Cathy, a certified clotheshorse, loved her house but was ready to giddyup and go when she ran out of closet space. That was, until her neighbor Dodie showed her the ropes about installing a closet system. Of course, the first rule of order was to muck out the closets.

Closet systems bring out the worst in people—that is, people who don't have them. When someone shows you her new closet system, you can't help but feel jealous. Especially if there's a slide-out shoe rack. A girl can only take so much!

But don't even think about throwing yourself a pity party, because we found a great closet system that you can put together yourself. And it comes in compact boxes so you don't have to pay to have it delivered. By schlepping and installing it yourself you can save at least 30 percent of the overall price, which means that you can put those extra dollars into buying another closet system, or more shoes . . . now that you have the extra space.

Purchasing the Closet System

We tried other brands of closet systems, and we can honestly say that the elfa system, which is sold through The Container Store, is by far the best one on the market. If you don't have a store in your area, then you can purchase a system online. However, if you can get yourself to a store, it's worth the trip because there you'll find elfa experts, who will help you design the closet as well as teach you how to install the system on a mock wall.

Before purchasing an elfa system, or any other closet organizer, you need to gather some information.

First, take exact measurements of the inside of the closet, including any obstructions (e.g., access panels, light switches and receptacles, etc.) and the height of the baseboards. The Container Store's computer will subtract 1 inch from the measurements for wiggle room.

Decide what your needs are for the closet. Think *visibility, flexibility, and accessibility*. Do you require a lot of short hanging for shirts? Long hanging for dresses and coats? Drawers at waist height? Shoe racks? Measure how much shelf and rod space is currently being used for each type.

What type of closet is it—a reach-in closet or a walk-in (more than 4 feet deep)?

Are the closet walls made of drywall, wood paneling, plaster, or concrete? All of this information is vital to getting the right system and hardware for your closet.

Once the closet has been designed and all the boxes and hardware have been pulled, go over every item with the elfa expert to make sure that it's all there. The last thing you want to do is to go home, get started, and realize that you're missing something.

Removing the Old Shelf and Clothes Rod

If the closet does not have a ceiling light, you'll need to set up a stand-alone flashlight and, if necessary, some light fixtures with extension cords.

In order for you to install the new closet system, you'll need to remove the old one. An ordinary closet typically has one clothes rod with a shelf above it. Look to see how these things are anchored to the wall. If you see screws, then use either a Phillips or a flathead screwdriver to remove them. If you see nails, then use the claw of a hammer to remove them. And if you don't see any fasteners, then unfortunately the rod and the shelf were probably attached with an epoxy adhesive. Just be grateful that the whole thing never came crashing down!

To remove a shelf and a clothes rod that are glued on, first lay down a floor covering. Next, place a metal putty knife at the edge of the shelf or rod bracket and hit the handle with a hammer. Continue until it becomes completely loose and then disassemble it. Use the putty knife to scrape away any loose paint chips and drywall.

Whenever you remove a clothes rod or a shelf from a wall, you're more than likely going to cause damage to the drywall, so be prepared to patch and paint the closet. And if you're thinking that the new closet system will cover the mess, stop to consider how much time and money you're investing in this project. It's better to take a little extra time now and do it right.

Getting Started

This is one of those projects that looks more intimidating than it actually is. In fact, there's just a small amount of construction involved—the rest is assembly.

Read the manufacturer's instructions carefully, familiarize yourself with the lingo (e.g., "hanging standard"), and have your personal closet design handy for reference.

Installing a Closet System

Note: The following directions are for a reach-in closet, which requires using one wall. For a walk-in closet, you will just be repeating this on all three walls.

Track Installation

Locate where you want to place the track across the wall. Use the hanging standard to determine the height (you want it to be above the baseboard), along with the personal closet design.

Standing on the stepladder, use the large level to mark a straight and level line across the wall.

Place the track on the wall so that the bottom of it is on the line. Holding the track with one hand, use the pencil

Marking track placement

to mark each of the holes in the track. (These holes will be used as the sites to drill in the anchors and screws.)

Using a ⅜-inch bit, drill holes into all the marked holes on the wall, avoiding ductwork, electrical wires, plumbing components, etc. Insert an anchor (pointed side toward wall) into a hole and pound it with the mallet until it is flush with the wall. Be careful not to get too carried away with hammering, because you'll need to remove the spacer, which is on the top of the anchor. Repeat with all the anchors and holes. If you're experiencing any difficulty inserting an anchor, then redrill the hole to clean the hole up a bit.

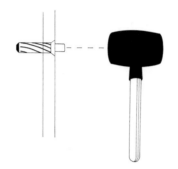

Inserting an anchor

Remove all the spacers and throw them away.

Place the track (notches on top) over the holes in the wall. Use the Phillips screwdriver to insert the screws just a bit so that they're semisecured.

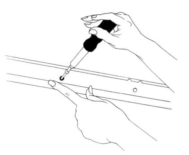

Installing track on wall

Now use a Phillips drill bit in the drill to insert the screws through the holes in the track and into the anchors. It's important to use the drill for this step because you need to insert the screws deep enough so that the anchors expand. Otherwise, the closet system will not be properly supported.

Note: It's really easy to strip a screw when you're installing it at eye level or above, rather than downward. When you're securing the screws and

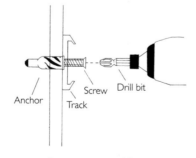

Anchor   Screw   Drill bit
Track

Inserting screws fully to expand anchors

you hear a clunk, clunk, clunk from the drill, that means that you're stripping the screw and you need to stop immediately. Put the drill in "reverse" to take the screw out just a bit, switch it to "forward," and drill it into the wood.

Use the Phillips screwdriver to hand-tighten all the screws.

Now that you have the track installed, the rest is just assembling. We told you it was easy!

Assembling Closet Shelves

Slide the hanging standards (vertical pieces) close to each other so that you're sure to install the brackets in the right bracket holes. Insert one bracket at a time into the bracket holes on the hanging standards.

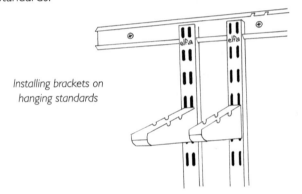

Installing brackets on hanging standards

Now install the shelf onto the brackets, snap into place, and then add the bracket covers over the sides to prevent the shelf from moving.

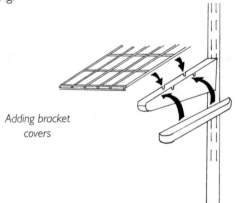

Adding bracket covers

Assembling Closet Clothes Rods

Slide the hanging standards (vertical pieces) close to each other so that you're sure to install the brackets in the right bracket holes. Insert one bracket at a time into the bracket holes on the hanging standards.

Hang the clothes rod holders in the underside of the brackets. Install the rod by placing it onto the rod holders and pushing down in the middle of the rod —don't use the ends to push down because you may get your fingers pinched.

Hanging rod holders

Assembling Closet Drawers

Slide the hanging standards (vertical pieces) close to each other so that you're sure to install the brackets in the right bracket holes. Insert one bracket at a time into the bracket holes on the hanging standards.

Separate the glider (U-shaped) from the frame. With the notches at the top of the glider, place the glider so that it's an open U.

Separating glider from frame

Snap the notches of the glider into place. Use the mallet to tap the protruding ends of the glider into position—this will secure the glider.

Snapping glider notches into place

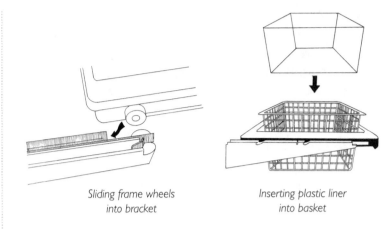

Sliding frame wheels into bracket

Inserting plastic liner into basket

Slide the frame wheels into the bracket (wheels to wheels) and push in. Drop a basket into place and attach the clips to keep the basket stationary. Insert a plastic liner (optional).

Tools Needed

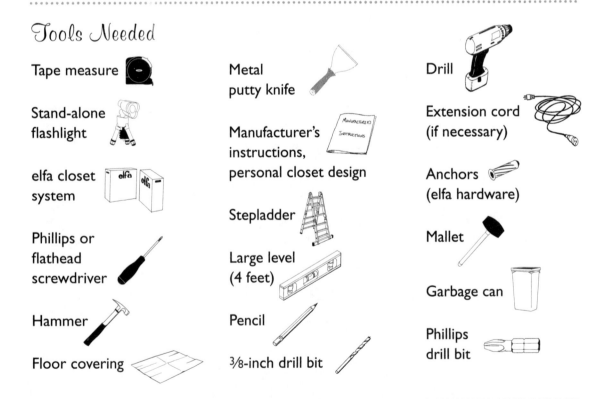

Tape measure

Stand-alone flashlight

elfa closet system

Phillips or flathead screwdriver

Hammer

Floor covering

Metal putty knife

Manufacturer's instructions, personal closet design

Stepladder

Large level (4 feet)

Pencil

⅜-inch drill bit

Drill

Extension cord (if necessary)

Anchors (elfa hardware)

Mallet

Garbage can

Phillips drill bit

Installing an Under-Sink Roll-Out Drawer

Margie had so much stuff under her kitchen sink that she didn't know about the leaking garbage disposal until she saw the water stain on the basement ceiling. Margie realized that if only she had organized things under the sink, she could have caught the leak. She also realized that she wouldn't have bought more **Brillo** pads since she already had three boxes.

Buying a Roll-Out Drawer

There's a lot going on behind the cabinet doors below the sink—the base of the sink, the sink trap, water lines, a hose from the dishwasher, and for most people, a garbage disposal. And that's not taking into account all the cleaners and other products you store there.

So, before heading out to buy a roll-out drawer, you need to know exactly how much space you have to work with. To do this, you'll need to remove everything (i.e., products, not plumbing) that's behind those doors. *Good times.*

Measure the narrowest width at the door opening of the cabinet, taking into account protruding hinges.

Now determine the depth of the cabinet by measuring from the back wall to the inside of the cabinet door. Be sure to allow enough room so that the door can close when the gliding drawer is installed.

Check where pipes or other obstructions are located and measure their distance from the cabinet walls and floor in order to determine the functionality of the sliding drawer.

Take these measurements to obtain the right size roll-out unit for your cabinet.

Getting Started

Now that the cabinet is empty, give it a good cleaning.

We recommend that you use a small stool or a pillow for comfort while doing this project. Also, set up a stand-alone flashlight so you can see inside the cabinet.

Spread a drop cloth near the cabinet and remove all the contents of the box containing the roll-out drawer. Read the manufacturer's instructions and check that all the parts are included.

Before you throw out the box, do a sanity check to make sure that you bought the right size. Temporarily place the drawer onto its retracted gliding frame and set it inside the cabinet to confirm.

Next, use the manufacturer's provided alignment guide (ours was cardboard and allowed a 2-inch front setback) to help ensure the cabinet door's clearance. Roll out the drawer completely to make

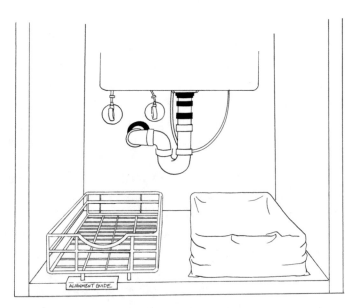

Confirming size and placement using alignment guide

sure that when it's fully extended, it provides the access you want. If not, adjust the frame within the limits that allow the cabinet door to close when the drawer is completely retracted.

Wait, you're not done testing it yet! Place all the items to be stored in the gliding drawer to be sure you're meeting your storage needs.

Okay, now you can install it.

Installing the Gliding Frame

Set up the stand-alone flashlight so that you can see inside the cabinet.

Retract the gliding frame of the drawer and place it on the cabinet floor. With a pencil, trace its front and sides onto the base of the cabinet floor.

It's important to make sure that the front frame marks are even. To do this, carefully measure from the front edge of the cabinet to the traced lines. Adjust the lines if needed to ensure that the drawer opens properly and not on an angle.

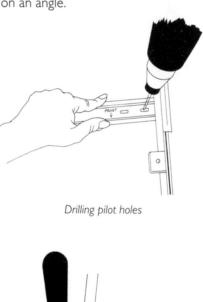

Drilling pilot holes

Set the gliding frame on the front mark and carefully align it with the side marks. Drill pilot holes (using ⅛-inch drill bit) into the holes in the front of the frame. Use an electric screwdriver or a Phillips screwdriver to insert the screws through the holes and into the floor of the cabinet. Once the front of the frame is secure, drill pilot holes in the rear frame holes and install the wood screws

Note: Wood screws are not made of wood, but rather are metal screws designed to grab into the wood for a secure fit.

Installing
wood
screws

Once the entire frame is secured, fully extend the drawer glides. Set the drawer on the glides, align the holes, and secure with provided mounting screws using the Phillips screwdriver. Do not overtighten.

Place your storage items in the drawer and glide it into the cabinet.

Securing drawer on glides with mounting screws

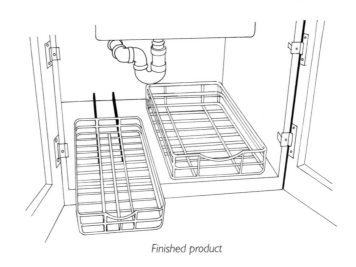

Finished product

Tools Needed

Tape measure

Pencil

Paper

Low stool or pillow

Stand-alone flashlight

Drop cloth

Drill

⅛-inch Phillips drill bit

Under-sink roll-out drawer

Manufacturer's instructions

Electric screwdriver or Phillips screwdriver

Home Safety

You've had a symbiotic relationship with your house ever since you first called it home, but now it's like you're total strangers. In fact, you're finding there's little to trust: the stairs are a bit too precarious, you can't seem to find your footing in the shower anymore, and even the front porch seems treacherous.

In this section, you'll learn things you can do to prevent accidents from happening inside and outside your home, from installing a shower grab bar to installing a motion-activated floodlight.

It's vital that you follow these safety measures, because calling your house "*home*" is one thing. Feeling *at home* in your house is another.

Installing a Shower Grab Bar

G eeta traveled a lot for her job, but the only sightseeing she ever did was *inside* hotels. The upside was that when it came time to remodel her bathroom, Geeta had seen so many that she knew exactly what she liked and what she didn't. The one must-have thing, besides a heated towel rack, was a shower grab bar. Now, if she could just find someone to turn down her sheets. Or better yet, place a chocolate on her pillow?

We know. We've done it, too. You're stepping into the tub and start to slip, so you grab onto the towel bar for support. That towel bar is designed to hold maybe five pounds, without force. *Enough said.*

There's no need to wait until you're injured or elderly to have a grab bar in your shower. Preventive steps, like this one, will keep you *in step*.

And speaking of steps, this project doesn't require too many, and none of them is difficult. You just need to get over your fear of drilling into tile.

Buying a Grab Bar

When you go to the home improvement store, you'll see safety bars and grab bars—you need to know the difference. Safety bars are non-permanent devices that fit onto a bathtub via suction. And because no drilling is required, the safety bar can be moved to a different location. The drawbacks are that most models cannot hold up to 250 pounds; they can adhere only to a bathtub, not a tiled wall; and they do not meet the Americans with Disabilities Act codes for people with disabilities.

That's why we recommend that you purchase and install a grab bar. Grab bars have come a long way in looks, so if you don't see something that appeals to you in the store, then go online. You'll be amazed at the different designs.

When choosing a grab bar, look for one that is at least 1½ inches in diameter (for children or a person with small hands, 1¼ inches is recommended), with 1½ inches of clearance from the wall. Some models are made of metal, and others come in colored plastic with a cushioned or scored section. Ask a clerk if you can take the grab bar out of the box so you can test drive it. Grab it to see how it feels—maybe you'll prefer the cushioned model to the round metal, or vice versa.

Select a grab bar that can hold at least 250 pounds (some go as high as 500 pounds) and has the ADA seal on its packaging, which means that the product has met the requirements of the Americans with Disabilities Act.

And last, purchase one with a concealed mount, because the screws will be concealed, and therefore, they won't rust. Plus, it just looks nicer.

Getting Started

Remove the parts from the box and make sure that everything is there. Read the manufacturer's instructions carefully and do any disassembling as required.

Take note of the size drill bits required for drilling into the tile and stud because manufacturers may differ.

Don't even think about installing a grab bar into the drywall. You have to install it into the studs that are behind the drywall, which means that you'll drill into the tile first, and then, using a different drill bit, you'll drill a little bit more into the stud. The grab bar has to be supported so that it can support you!

One more thing. You need to decide how you want to position the grab bar—vertically, horizontally, or at a 45-degree angle. If more than one person will be using the grab bar, you may want to consider angling it, because that makes it easier for people of different heights to use.

Locating the Studs

As we stated above, you must attach the grab bar to studs, which means that you need to locate the studs that are behind the wall. This is easy to do, unless the wall is completely covered with tile, because stud finders don't work on tile.

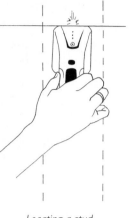

If the Wall Is Partially Tiled

Place the stud finder above the tiles on the drywall, moving it across until you see a light, hear a tone, see an LDC display, or all of the above. Note the site with a small pencil marking. Find the next stud, which should be 16 inches away. Mark that site, too.

Locating a stud

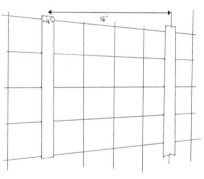

Starting at a pencil marking, run a long piece of masking tape (1½ inches wide) all the way down the tile, keeping it straight with the help of the 4-foot level. Repeat at the other marking(s).

Marking length of stud

If the Wall Is Completely Tiled

Because a stud finder won't work on tile, you'll need to do a little bit of searching with a little tiny masonry drill bit (⅛ inch).

Using a stepladder if necessary, start at the top row of tiles, measure out 16 inches from the edge and note the location on the tile with a small pencil mark. Wearing safety glasses and dust mask, drill a small hole into the grout that's at or near the mark. If the drill bit hits wood—*eureka!* That's the stud. If not, then insert a thin piece of wire into the hole and move it around until it touches the stud. To find the next stud (which should be 16 inches away), drill into the grout. Mark that site, too.

Starting at a pencil marking, run a long piece of masking tape all the way down the tile, keeping it straight with the help of the 4-foot level. Repeat at the other marking.

Installing the Grab Bar

If you've decided to attach the grab bar horizontally, then measure about 33 to 35 inches above the floor of the *bathtub* for the optimal height. Mark the measurement onto the pieces of tape. For vertical and angled positioning, the height is your preference. (We chose to angle the grab bar.)

Place the grab bar in the desired position and completely pencil in the 2 mounting holes onto the 2 pieces of tape. Be careful not to move the grab bar while you're doing this.

Remove the grab bar. You should now see 2 circles identical in size. This is where you'll be drilling.

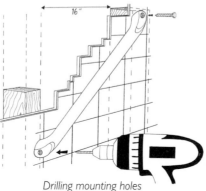

Pencil in mounting holes

Insert the carbide drill bit into the drill (we used a $5/16$ inch).

Wearing the safety glasses and dust mask, place the drill bit onto the circle on the tape. Starting slowly so that the drill bit doesn't drift away, drill firmly into the tile. It will only take a few seconds before you hit the stud. Immediately stop. Repeat on the other tile.

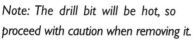

Drilling mounting holes

Note: The drill bit will be hot, so proceed with caution when removing it.

Now replace the carbide bit with the smaller bit (we used a $3/16$ inch) and drill through the hole in the tile into the stud. Repeat on the other tile.

Remove the masking tape.

Position the grab bar so that the holes align over the holes in the tile. Place the washer onto the lag screw and then insert it through the grab bar into the hole.

Use the socket wrench to tighten the screw, being careful not to overtighten to avoid cracking the tile. Repeat on the other tile.

Snap the covers onto the grab bar, if necessary.

If you drilled into the grout to find the stud, place a dab of silicone on the tip of the wire and apply it to the hole(s).

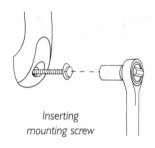

Inserting mounting screw

Tools Needed

Grab bar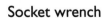

Manufacturer's instructions

Stepladder (if necessary)

Stud finder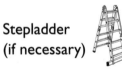

Pencil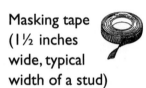

Masking tape (1½ inches wide, typical width of a stud)

4-foot level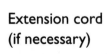

Drill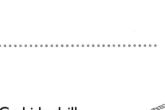

Masonry drill bit (⅛ inch)

Extension cord (if necessary)

Safety glasses

Dust mask

Thin piece of wire (if necessary)

Carbide drill bit (check with manufacturer for size)

Drill bit (check with manufacturer for size)

Socket wrench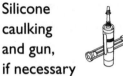

Silicone caulking and gun, if necessary

Installing a Deadbolt Lock

Kelsey locked herself out of the house and was in dire need of getting back in. While she was searching the lower windows of the house for an easy entry, her neighbor Rebecca stopped by to see if she could help. With credit card in hand, Rebecca slipped the card between the door and doorjamb, and voilà, the door opened. "American Express ... don't leave home without it," Rebecca quipped.

A door that doesn't have a deadbolt lock is no laughing matter. It took seconds for Rebecca to break into Kelsey's house.

A deadbolt lock can actually add ten to fifteen minutes to the time it takes for someone to break in through a door, making the effort enough of a hassle to send the thief elsewhere.

Buying a Deadbolt Lock

When buying a deadbolt, you should consider the color and style of the existing doorknob (a.k.a. handset). And if you're putting a deadbolt on an interior door, you need to be sure that the deadbolt will work with the thickness of the door, because most are designed for exterior doors, which are thicker. Look for one that has an "adapter ring" inside the lock mechanism that can be removed to use on a thinner door.

We highly recommend purchasing a door lock installation kit, which includes 2⅛-inch and

Deadbolt lock

1-inch hole saws and a mandrel (a drill bit attachable to both hole saws), especially one that has a guide for drilling, a chisel, and a hinge template, too. These kits are fairly inexpensive and make the job much easier.

Getting Started

This project will create a lot of sawdust, so be sure to lay a drop cloth underneath the door.

Determining the Correct Placement

Deadbolts are normally installed above the doorknob. In fact, the industry standard is 4 inches between the center of the doorknob and the center of the deadbolt. But depending on the architecture of your door and the size of the doorknob and lock, you may want them farther apart.

Have your helpful friend place the deadbolt lock against the door so you can choose the optimum height for it.

Note: You should do this on the interior, not the exterior, side of the door.

Use a pencil to mark the site.

Always begin working from the interior (i.e., the side where the hinges are located).

Deadbolts (and doorknobs) are normally installed either 2⅜ inches or 2¾ inches from the edge of the door. This distance is called the "backset." Both the

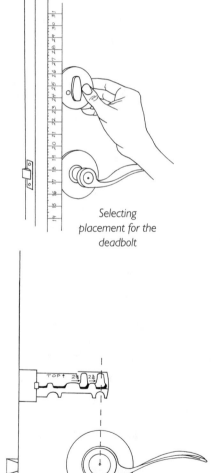

Selecting placement for the deadbolt

Backset measurement choices

lock sets and templates normally allow for either of these distances. To determine which backset you'll be using, measure the distance from the center of the doorknob to the edge of the door.

Using a Door Lock Guide

Once you determine the correct backset, adjust the door lock guide to the correct setting, according to the manufacturer's instructions. Place the guide so the mark is centered in the guide's largest hole and flush against the door. Clamp the guide in place—some may have a locking tab on top to do this.

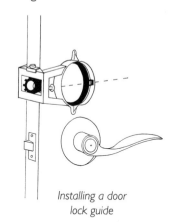

Installing a door lock guide

Boring a Door Face Hole

Insert the provided mandrel (a.k.a. drill bit) into the 2⅛-inch hole saw and attach to the drill.

With the door open, place the drill in the guide.

Note: Only the mandrel will touch at first. Check to be sure that the mandrel's tip is on the mark—you can do this by pushing the tip into the door and then removing it to see if a slight mark is left.

Hold the door steady by placing your foot against it. Gently but firmly pull it toward you. Drill into the door until the mandrel shows through on the other side. Remove the hole saw from the hole in the door.

Boring a door face hole

Note: Do not put the drill in reverse to remove it; continue to drill but pull back slowly and carefully. Reversing the drill may loosen the bit and leave it in the door as you pull back.

Approach the door from the exterior side and insert the mandrel's tip into the hole. You'll be finishing the drilling on this side, because in order to get a clean hole, you need to drill on both sides of the door. Once the 2⅜-inch hole is made, carefully remove the plug (piece of wood) from the saw bit, being careful because the bit may be hot.

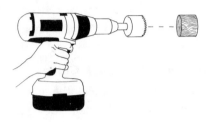

Removing bored wood from saw bit

Boring a Latch Hole

Allow the mandrel to cool and then move it to the 1-inch hole saw. Double-check the guide on the door to make sure that it's completely tightened and aligned with the hole that you just drilled.

Using the guide, drill a 1-inch hole through the edge of the door. This hole, called the "latch bore hole," will allow the deadbolt's locking bolt to pass through the door and into the jamb. This hole should go through the short edge of the door into the large hole previously drilled. Remove the hole saw.

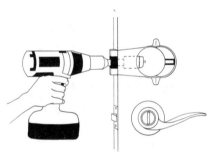

Boring a latch hole

Installing the Faceplate

Many deadbolt locks have two faceplates for the bolt: a rectangular one and one with rounded corners. It's up to you to decide which to use. Install the faceplate onto the bolt and then place it into the latch bore hole. Use a pencil to trace the faceplate and its screw

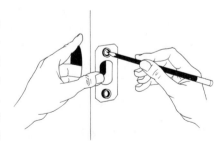

Tracing the face plate and screw holes

holes onto the edge of the door. Using a pilot drill bit, drill pilot holes for the faceplate screws.

You'll need to remove the wood within the traced outline so that the faceplate fits flat and evenly with the edge of the door. To do this, place the chisel against the line. Be sure the beveled (angled) side is inside the area to be removed. Firmly hit the chisel with a hammer, being

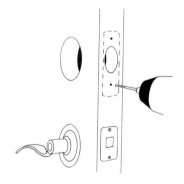

Drilling faceplate pilot holes

careful not to hit it too hard. Repeat this around the outline to score all the edges. If you're using a rounded faceplate, the process is a bit more difficult because you'll need to use the corner of the chisel to score the curves.

Now place the chisel along the scored line with the beveled tip facing down, against the wood to be removed. Hold the chisel at an angle and gently but firmly tap it with the hammer. Adjust the angle to be sure that you don't go too deeply and that you take off only a little at a time. Test your work by setting the faceplate in place; if more wood needs to be removed, repeat the process.

Once the faceplate fits flush against the edge of the door, install the bolt assembly with the bolt retracted. To do this, place the bolt assembly through the hole in the edge of the door so that it extends into the larger hole. Be certain that the correct side of the bolt

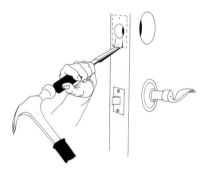

Chiseling door edge

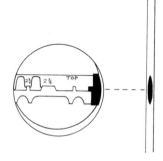

Installing bolt assembly

assembly is facing upward (check manufacturer's instructions). Using a Phillips screwdriver, insert the screws into the faceplate.

Assembling the Lock Mechanism

Assemble the lock mechanism by carefully aligning the spindles with the bolt assembly. Next, on the interior side of the door, attach the other half of the lock mechanism by lining up the two screw holes. Use a Phillips screwdriver to insert the long screws. Test to be sure the bolt opens and closes smoothly—you may need to move it around a bit.

Aligning the Door Jamb

Manufacturers recommend tracing through the latch hole on the doorjamb to mark the strike box location. We discovered that the best tool for this job is lipstick (or chalk). Cover the bolt face with lipstick. Close the door and then turn the deadbolt latch. Lipstick will transfer onto the doorjamb and mark precisely where the opening of the strike plate should be.

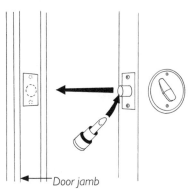

Door jamb

Marking strike plate opening

Set the strike plate on the doorjamb, being sure the lipstick marking is showing through the open hole. Make sure that the faceplate is straight and then trace all the edges with a pencil and mark the holes.

Drill pilot holes for the screws. Open the door and fully extend the deadbolt. Measure the bolt's length and transfer that measurement in pencil on the I-inch hole saw as a depth guide to indicate where the drilling should stop.

Marking drill's depth guide

Using the 1-inch hole saw, drill the center hole for the deadbolt latch. Next, working as you did on the door edge, use the chisel to score the edges for the strike plate and remove excess wood so that it fits flush with the doorjamb.

Use the Mr. Clean Magic Eraser to remove all remaining pencil marks and lipstick. Vacuum the dust and wood shavings from the area.

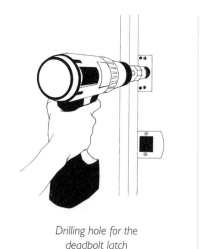

Drilling hole for the deadbolt latch

Tools Needed

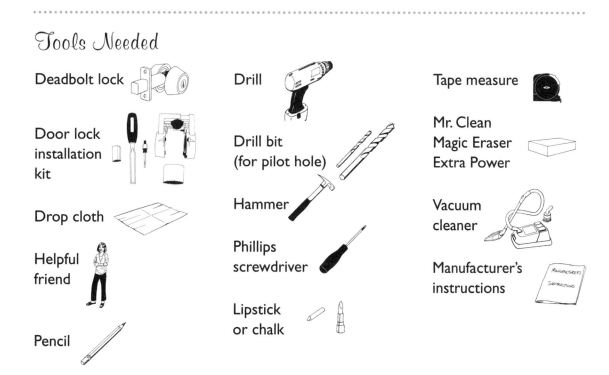

Deadbolt lock

Door lock installation kit

Drop cloth

Helpful friend

Pencil

Drill

Drill bit (for pilot hole)

Hammer

Phillips screwdriver

Lipstick or chalk

Tape measure

Mr. Clean Magic Eraser Extra Power

Vacuum cleaner

Manufacturer's instructions

Installing an Extra Stair Handrail

Hire an electrician to install lighting at the bottom and top of the stairs.

Chris knew that an extra handrail would be a good safety device to install in her mother's home. What she didn't count on was how much she would be using it while visiting. So, she installed an extra handrail in her own house. "Safety devices aren't for just the young and old," Chris decided. "They're for everyone in between, too."

The number one cause of injuries in homes is falls, so make installing an extra handrail your number one priority. We want you to fall head over heels for stair safety . . . figuratively speaking, that is.

Every interior staircase has a handrail, but the Home Safety Council recommends that every staircase have two. And we agree. An extra handrail provides you with extra support going up or down a flight of stairs, *no matter your age.*

Buying a Handrail

Before you go to the store, contact your county government to find out the building code for installing an interior handrail.

Measure the length and depth of the existing rail. Note any details in the wood or, better yet, take a photo of it with your camera or cell phone, so that you can find a match.

Another thing to note is how many brackets are used for the existing rail (typically one for every 48 inches) and the color (typically brass).

All the handrails in home improvement stores come unfinished. You may be able to find a store that can custom-order a prefinished model for you, but expect to pay a lot more for it. We recommend that you buy one unfinished and stain or paint it yourself, to save money and get a perfect match.

You also probably won't be able to purchase the handrail in the length that you want, but depending on where you shop, the home improvement store may cut it to the desired length for you. Just know that they do only straight cuts, not angled.

Getting Started

Use the tape measure to find the distance from the edge of the first step to the top of the existing handrail, and the edge of the last step to the top of the handrail. Now use the tape measure to locate these heights at the top of the first and last steps along the opposite wall and mark them with a pencil.

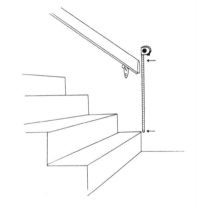

Measuring the existing handrail

Now that you know the exact height that you'll be installing the extra handrail, you need to find the studs to support it. Run the stud finder across the wall, and mark the beginning and end points of the stud with a pencil line. You should have one bracket located at least every 48 inches. And you'll need 1 stud for every bracket.

Flip the rail so that it's upside down. With the assistance of your helpful friend, hold the rail against the wall at the marked height. Starting at the top of the rail at one end, run a light pencil line on the wall, going the entire length of the rail. Remove the handrail. This line marks where the top of the rail will rest.

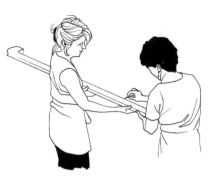

Marking new handrail location

Measure the depth of the new handrail (ours was 2½ inches). Now place the tape measure on the line (the one you just made) and at all the stud locations, go down 2½ inches, and mark the spot with a light pencil line. These lines indicate the bottom of the handrail, where the brackets will attach to the handrail.

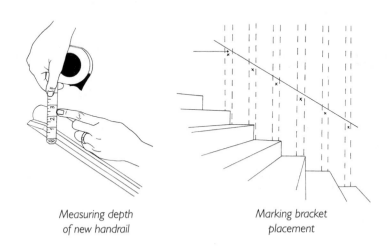

Measuring depth of new handrail

Marking bracket placement

We chose to use a Phillips screwdriver instead of an electric drill because brass screws strip easily.

Installing the Extra Handrail

The first step is to attach the brackets. The brackets have to be installed just *below* the bottom of the rail and into a stud. So, hold a bracket against the wall so that the top part of the bracket is just below the line that indicates the bottom of the handrail. Now pencil in the bracket holes. Repeat this procedure with the other brackets.

Using the drill bit recommended by the manufacturer, drill pilot holes into the marked holes on the wall. Position the bracket so that its holes align over the pilot holes. Insert the screws with a Phillips screwdriver or drill.

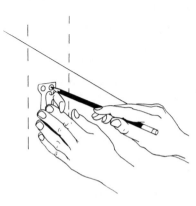

Marking bracket holes

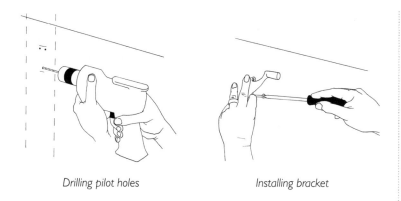

Drilling pilot holes   *Installing bracket*

Note: If you're having difficulty inserting the screws, you may need to drill a slightly larger pilot hole.

Lay the rail on top of the brackets and have your helpful friend hold it still. Place a strap over a bracket and pencil in the 2 bracket holes onto the bottom of the rail. Place the rail on the floor.

Using the drill bit recommended by the manufacturer, drill pilot holes into the marked holes on the bottom of the rail. Place a strap over the holes and insert 1 of the screws through it and into the rail. If it's too challenging to insert the screw, remove it and make slightly larger pilot holes.

Note: Do not completely attach the strap. You want it hanging by a screw.

Lay the rail on top of the brackets. Place the hanging strap over a bracket and insert the other screw(s). Have your helpful friend hold the rail in place.

Attaching strap over a bracket

Now that it's secured in one section, go to the next bracket, place a strap over it, mark its holes, drill pilot holes, and insert the screws. Repeat on the other bracket(s).

To make it easier to insert a screw into wood, rub the threads of the screw with a bar of soap.

Once you're done, go back and hand-tighten all the brackets, and use the Magic Eraser to remove all the pencil lines, being sure to test it on a hidden area first to avoid damaging the painted wall.

Tools Needed

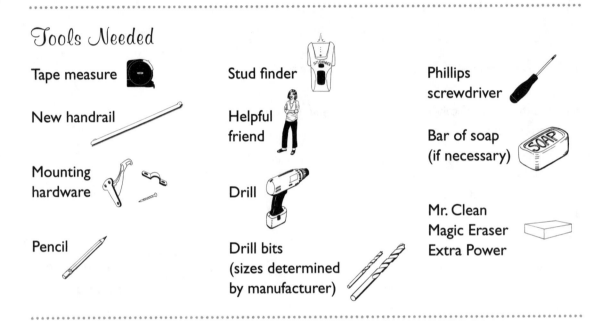

Tape measure

New handrail

Mounting
hardware

Pencil

Stud finder

Helpful
friend

Drill

Drill bits
(sizes determined
by manufacturer)

Phillips
screwdriver

Bar of soap
(if necessary)

Mr. Clean
Magic Eraser
Extra Power

Installing a Motion-Activated Security Floodlight

Motion-activated lighting is a great tool for letting strangers know that even though you may be sleeping, your house is *wide-awake*!

Exterior motion-activated lighting is one of the least expensive ways to add a layer of security to your home. It's also a great way to easily access light on demand, such as when you realize at 11:00 p.m. that you forgot to take the trash out.

You may think that this light fixture works just by sensing motion, but it also works by sensing heat from people, large animals, and automobiles, too. So be sure to install it in the right location or the neighbors across from you may never get any sleep!

Buying the Light Fixture

Motion-activated lighting has what's called a detection zone. This is the area that is covered by the motion/heat sensor of the light fixture.

If you're shopping in the exterior lighting aisle of a home improvement store, you'll find lots of products that are motion activated, but they're mainly designed for front door use and therefore have a small detection zone.

But intruders don't usually knock on the front door.

To increase the security of your home, you need to provide lighting to a greater area, not just the front porch. Therefore, we recommend purchasing security lighting (a.k.a. floodlight), because it has a larger detection zone.

Be sure to look on the box for the words "security" and "flood-light," and also for the detection zone (e.g., 180 degrees). It's also important to find the manufacturer's suggested height for installation of the product (most prefer the floodlight to be installed 8 to 12 feet above the ground).

Getting Started

Flip on the switch to the exterior light. Turn off the power to the exterior light fixture at the main service panel. Have your helpful friend tell you when the light is off.

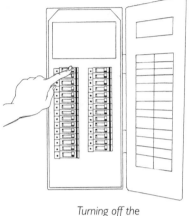

Turning off the electric power

Use the voltage tester to see if the power was turned off to each of the lights. If the tester emits a constant beeping noise with a flashing light, that means the power was not turned off. Keep flipping the circuit breakers until the electricity has been turned off. Then mark the circuit breaker map accordingly.

Removing the Old Light Fixture

With your fingers, loosen the set screws that secure the glass housing. Remove the glass

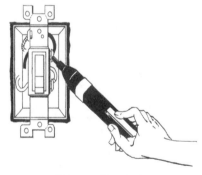

Using a voltage tester

housing and the lightbulb and set them in a safe place.

Remove the old light fixture by loosening the screws or decorative nuts (a.k.a. acorn nuts), and put the screws in a secure place (e.g., your pocket). Wipe away any dust or cobwebs.

Once the light fixture has been moved away from the exterior wall, you'll be able to see the wiring.

Take a look at how the wires are connected—they should be black to black, white to white, and green or bare copper to green or bare copper. But that's not always the case. The rule of thumb is that since the wiring works, you'll have to reattach the wires from the new light fixture to the house wires the exact same way. So, draw on a piece of paper exactly how the wires are connected.

Detaching the Old Wiring

What you'll most likely see is a white wire from the junction box attached to the white wire of the old light fixture; a black wire from the junction box attached to the black wire of the old light fixture; and a grounding wire (green or bare copper) from the junction box attached to the grounding wire (green or bare copper) of the old light fixture. Each group of wires is twisted together and connected with a wire nut on top.

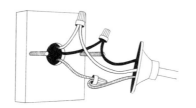

Detaching old fixture

Note: If you don't see a green or bare copper house wire, then you need to stop. All new light fixtures are manufactured with a grounding wire, which needs to connect to the home's grounding wire. So, button up the light fixture and hire an electrician to have a grounding wire installed to that box, as well as to check the home's entire electrical system.

Twist off all three of the wire nuts and loosen the wires. The old light fixture is now completely disconnected. Hand it to your helpful friend. Almost every new light fixture will fit onto the old mounting plate. Save the new mounting plate and hardware for possible future use.

Installing the New Security Lighting

Note: The light fixture will come with the bulb attached.

If the old light fixture works, don't throw it out. Recycle it instead by donating it to a thrift store or a Habitat for Humanity ReStore in your area.

If you're nervous about stripping a wire, buy a piece of wire from the hardware store to practice.

Attaching the New Wiring

Check that there's at least ½ inch of exposed wiring on each wire so that you'll get a good connection. If more wire needs to be exposed, use the wire stripper. Simply place the end of the wire into the proper hole of the wire stripper and the proper length that you want stripped. Clamp down and pull off the insulation.

Place the gasket (round ring) over the mounting hole. It should stay on without assistance. If not, place it on the back of the light fixture with the wires running through it.

If you have a helpful friend, have her hold the security light while you attach the wires. You can do this on your own, but it's just easier with an extra pair of hands.

It's always best to connect the grounding wires first, so twist them together and secure with a new wire nut. Finger-tighten. Now twist the white wires together and secure with a new wire nut. Finger-tighten. And last but not least, twist the black wires together and secure with a new wire nut. Finger-tighten. Push the wires into the junction box.

Installing the Security Light

Insert the screws (provided by the manufacturer) into the 2 holes

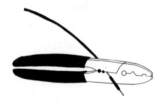

Stripping wire

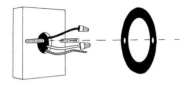

Placing gasket over mounting hole

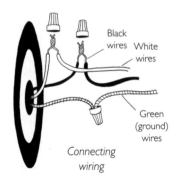

Black wires White wires

Green (ground) wires

Connecting wiring

Securing cover plate

on the security light's cover plate. Position the plate so that the screws are aligned with the holes in the junction box. Insert the nut and use the adjustable wrench to tighten.

Sealing the Security Light

Before turning on the power to the security light, apply a thin coat of silicone caulk (following the light manufacturer's recommendation) around the edge of the cover plate to prevent any moisture from entering.

Turning on the Power

Turn on the power at the main service panel and at the switch to the light.

Adjusting the Timing

Refer to the manufacturer's directions for adjusting the security light to the desired timing.

When it's time to replace the halogen bulb, be sure not to touch the bulb with your bare fingers, because the oil from your skin can shorten the life of the bulb. Instead, use a soft cloth or wear gloves.

Tools Needed

Voltage tester

Ladder (if necessary)

Paper and pencil

Adjustable wrench

Motion-activated security halogen floodlight

Helpful friend

Wire stripper (if necessary)

Silicone caulk (if necessary)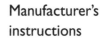

Manufacturer's instructions

Outdoors

The doormat may say "Welcome," but is your house screaming, "Go Away"?

It's hard to check your home's curb appeal if you never use the front door. So, walk across the street to your neighbor's house and then turn around and take a good look at yours. Do you like what you see? Do you like what *she* sees?

In this section, you'll learn how to safely remove mold from the siding, clean clay off the walkway, and fix a wood fence—and that's just for the front of the house.

Of course, there is another option . . . your guests can enter through the garage. *Yeah, we didn't think so.*

Maintaining Siding, Decks, and Walkways

Does your house need a facelift? Before you start getting quotes for new siding, decking, and walkway, give them all a good washing. If it were only that simple for us!

When you bought your house, you believed that you cleaned the inside and Mother Nature's rains would take care of the outside. Who knew that the house needed soap to go with those downpours?

Leaving the outside cleaning to rainstorms is one of the biggest mistakes homeowners make. That's why gutters get clogged, mildew forms on siding and decking, and clay discolors the concrete.

We know that you don't want to add any more areas of cleaning to your already too-big list, but exterior cleaning has to be done only twice a year. Just don't let your son know that schedule!

Cleaning Siding

No matter which type of siding you have—aluminum or vinyl—it needs to be cleaned, even if you don't see any mildew. Why? Just do one section and you'll understand. Instead of rain cleaning siding, it can actually bring with it dirt, pollen, and residue from surrounding trees. We promise you that once you clean the siding, you'll be amazed at how much better your house looks.

Manufacturers of vinyl and aluminum siding will tell you not to use a pressure washer because the pressure can damage the integrity of the siding as well as force the water underneath the siding so that the wood is damaged. Also, the warranty for the siding may be nulli-

When purchasing cleaning products, think green. They may take a little more elbow grease from you, but that kind of grease is environmentally friendly!

fied if a power sprayer is used. So, how should you clean the siding? The old-fashioned way . . . with soap and water.

Getting Started

This job is best done on a warm sunny day. And we strongly recommend that you wear clothing that you won't mind getting wet.

Close all windows and doors!

Place the drop cloths over any bushes or flowers that border your house to protect them from soapy residue.

Cut one plastic bag into pieces large enough to cover any exposed exterior electrical outlets and secure with painter's tape.

If necessary, set up a ladder near the house, using the appropriate safety measures.

Washing the siding is just like washing your car . . . you'll start at the top, working in sections, hosing down the area first, cleaning it, and then rinsing.

Set the nozzle to a high-powered force and spray the first section at the top of the house. Place the extension wand (with the

Your home may actually have both vinyl and aluminum siding. Vinyl siding will feel like heavy plastic. Aluminum siding, which has a tin feeling to it, is more pliable than vinyl and is often used for wrapping exterior wood trim.

Tools Needed

Plastic drop cloths

Plastic bags

Scissors

Painter's tape

Garden hose with spray nozzle

Extension wand with soft brush attachment

Ladder (if necessary)

Bucket

Dish detergent

Invest in a good-quality spray nozzle for your hose. It will make the job easier.

Cleaning siding

brush) into the bucket of soapy water and clean the siding. Once you've reached the bottom of the section, place the brush back into the bucket and spray the area to wash away the soap.

Repeat these steps, working all the way around your house. When finished, remove the drop cloths from the shrubbery and plants, and the plastic covering the electrical outlets and lighting.

Cleaning a Deck

The typical life span of a wood deck is about twenty years. Of course, under extreme conditions (heat, direct sun, snow), it could be less. The best way to extend the life of your deck is to properly maintain it, which means cleaning it at least once a year, fastening nails and/or screws, and replacing any rotted boards.

Buying a Deck Cleaner

Before buying a deck cleaner, you need to know the size of your deck. So use the tape measure to determine the size of the area that you'll need to clean. And don't forget to include the stairs and railings, too. This really is important because you don't want to be cleaning the deck only to run out of product halfway through. So, buy extra because you can probably return an unopened container with a receipt.

There are so many different kinds of deck cleaners and applicators on the market that sometimes the choice will come down to the size of the deck and how much time you have to clean it. If you have a small deck, then we recommend using a deck-cleaning product with a deck brush.

However, if the deck is large and you don't have a lot of time, then we recommend that you either purchase or rent a pressure washer (you might be able to share the cost and product with a neighbor).

There are gas and electric models, and of the two we recommend the electric, because it's not as heavy.

You may already know that a pressure washer uses water to remove stain, dirt, and mildew. But when a deck cleaner is added to the water, it can clean the deck while reducing the time it takes to get the job done. Now we have your attention, right?

Do not use a pressure washer on a painted deck, because it will remove the paint.

Tools Needed

Tape measure

Deck cleaner

Pressure washer

Outdoor extension cord (if necessary)

Manufacturer's instructions

Safety glasses

Garden hose with spray nozzle

Deck brush

Getting Started

This project is best done on a warm, sunny day so that the deck dries quickly.

We know that you read the manufacturers' instructions before buying the cleaning product and the pressure washer, but we want you to read them again before beginning. We're funny that way.

Remove everything from the deck and sweep off as much dirt as possible. Hose down nearby flowers, shrubbery, and grass to protect them from the solvent.

Close all windows and doors that are near the deck.

Cleaning the Deck

Once the pressure washer with cleaner is ready to go, set it to a low speed, and spray a small section of the deck to test it, as well as to get a feel for how *powerful* the *power* washer can be.

Wearing safety glasses, start at the part of the deck farthest from the stairs and spray the wood, moving from side to side, until the area is completely covered. You should be able to immediately see the dirt, mildew, and old stains lift from the wood.

Be sure to clean the stairs and both sides of the railing.

Using a pressure washer

Spray the entire deck with water to remove any residue. If there is a persistent stain, use the deck brush to scrub it and then use the power washer on the area again. Spray it with water.

Cleaning a Concrete Walkway and Porch

If you live in an area that has red clay, you know all too well how it can stain concrete. Even regular old dirt seems to just soak into it. Here's a simple way to stay on top of the problem.

Buying Cleaner

There are products that will clean the concrete to rid it of dirt, and then there are products that will clean it *and* take some of the first layer off, too, which is called "etching." This method is used to prepare the concrete for paints, stains, etc. So, when purchasing a cleaner for a concrete walkway, be very, very careful not to buy a product that has the word "etcher" on the label.

As with other cleaning products for decks, there are a lot of choices for cleaning concrete, too. We found that TSP-PF detergent, which is phosphate free, worked well and was not as harsh as some other products.

Getting Started

We know that you read the manufacturer's instructions before buying the product, but we want you to read them again before beginning. We're funny that way.

Tools Needed

TSP-PF detergent

Garden hose with spray nozzle

Bucket

Deck brush or broom

Cleaning a concrete walkway

Remove everything from the walkway and porch, if necessary, and sweep off as much dirt as possible. Hose down nearby flowers, shrubbery, and grass to protect them from the solvent.

Close all windows and doors that are near the walkway and porch.

Cleaning the Concrete

Hose down the walkway. Following the manufacturer's instructions, combine the suggested amount of water and TSP-PF in a bucket.

Starting at one end, pour some of the mixture onto the walkway and use the push broom to scrub it clean. Continue the process the length of the walkway.

Cleaning a Brick Walkway

Follow the same instructions for cleaning concrete.

Repairing a Wood Deck

*T*he signs were all there from years of extreme exposure to the elements—dryness, cracking, rapid aging. Does it make you feel better that we're just talking about your deck?

A deck is the Rodney Dangerfield portion of a house—it gets no respect. In the summer, it's sunbathing without any sunblock; in the fall it's getting stained with fallen leaves; and in winter, it's holding up piles of snow because you're too busy shoveling the front. Yet, what's the first place you think about when considering where to host a party, huh? It's time to give the deck its much-deserved TLC.

The most common problems for a wood deck are loose boards, deep cracks, knots that have turned into holes, and squishy boards. No matter the problem, there's no need to throw the baby out with the bathwater. You can just replace a single board.

Note: If the deck has never been inspected by your county, do not do any repairs to it. It is imperative to your safety and the safety of anyone who sets foot on the deck that it get inspected and approved by your county's inspector before you proceed.

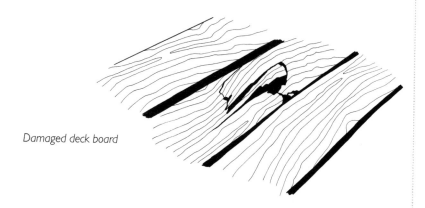

Damaged deck board

Buying Deck Wood and Screws

You need to know the height and width of the deck board(s) you'll be replacing. Don't panic if the wood in the store is wider than the boards on your fence. When a board is cut at the lumberyard, it is cut to a specific size. Yet, by the time it arrives in a store, it has shrunk, and after you install it, it will shrink again. That's because the wood contains water, and the water evaporates over time.

When replacing a piece of deck wood, you can upgrade to a better quality of wood (you pay more for fewer knots), but you have to keep the thickness and type of wood the same.

Before heading to the home improvement store or lumberyard, you need to know the measurement of the board(s) and how many you'll need. First, slowly walk up and down the deck and, using a piece of chalk, mark any boards that are very squishy to the touch of your foot, have deep cracks, and/or have large knots that have turned into holes. If any board has just one of these problems, it will need to be replaced. Using a long tape measure, measure the length of the board you'll be replacing (all the boards should be the same length).

When you're in the deck wood aisle, you'll see the boards stacked in sections according to size and price. The higher the price the better the wood, but even expensive boards can have knots and be warped. So, remove a board and prop it on its edge so that you can run your eye down its length to see if it's warped. If so, try another one. And then look at both sides to make sure it's blemish free.

You can have a board cut to the exact measurement at the home improvement store or lumberyard, but they only do straight cuts, not diagonal.

If your car has a roof rack, have a store clerk secure the board tightly with rope. If the piece is small enough and your car is big enough, you can slide it inside. If any part of the board extends out the back window, tie a piece of white cloth to it to flag its length to other motorists.

Don't forget the screws! The best fastener for securing deck wood is a self-tapping galvanized wood screw, typically 2½ inches.

Self-tapping screws have a very sharp tip, which prevents the wood from splitting, kind of like drilling a pilot hole. The galvanized part means that the screw won't rust.

You'll probably just need a handful of screws per board, but it's best to buy a box of them. Also, make sure that you have a drill and a Phillips drill bit for this job.

There's one other tool you may want to purchase, especially if the fasteners (e.g., screws and nails) are rusted, and that's a reciprocating saw. This baby can cut through wood and metal and save you a lot of time in the process.

Note: Using this saw is a rush. We promise that you'll want to find other things around the home for which you can use it.

Securing a Loose Board

If a board is loose, chances are that it was secured with a nail instead of a wood screw. A nail can pop up, but a screw has threads that dig into the wood, making it difficult to move.

Wearing the safety glasses, remove the nail that's popped up with the claw of a hammer. If the nail won't budge, try using a cat claw to pry it up. And if that doesn't work, hammer it into the wood. Then, using a drill with a Phillips drill bit, insert a self-tapping galvanized wood screw (2½ inches) next to the nail.

Removing a nail

Replacing a Deck Board

The first step is to locate the fasteners that connect the damaged deck board to the joists and remove them with a screwdriver or the claw of the hammer. Sometimes you'll see that the builder just used whatever fastener was in his nail apron, so you may have a mixture of both screws and nails.

The reality of working with wood is that the fasteners may be in too deep or are completely rusted, making them impossible to budge on the first, second, or third try. Since you're replacing the wood, don't be afraid to gouge the wood around the nail or screw with a Phillips screwdriver. The wood will tear away enough to provide you with better access to the fastener.

If you're having difficulty removing a nail, use a cat claw with the hammer (you have your safety glasses on, right?). This works just about every time. If the nail or screw isn't budging, you can use the Phillips screwdriver to gouge around it so that you can get more leverage. Don't worry about damaging the wood, because you're removing it, remember? So dig in, if you must!

And if all else fails, you can use a reciprocating saw. Wearing safety glasses, cut into the board a few inches away from the location of the screws and nails. The reason is that these are fastened into the joist, and you don't want to cut into the joist.

Using a reciprocating saw

Once you've removed a piece of board, you can choose to pry up the other pieces with the claw of the hammer or use the saw to cut them up.

If any nails are still in the board, hammer them down to prevent any injuries.

Place the board (or pieces) into a large garbage can.

Installing a New Deck Board

Now that the board has been removed, you'll see the joists in plain view. You'll also see debris that's collected on them. This needs to be removed before installing the new board so that it lies flat.

Lay the deck board on top of the joists. You'll be inserting 2 screws through the deck board and into each joist. And because you're using self-tapping deck screws, you won't need to create pilot holes.

Put on the safety glasses. Using the Phillips drill bit, insert the screws through the board and into the joists.

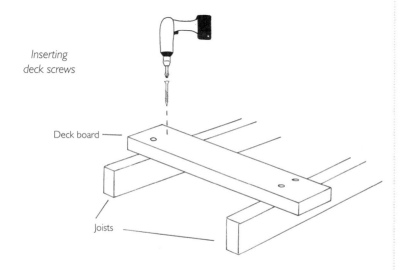

Inserting deck screws

Deck board

Joists

Note: You can strip a screw by using the wrong size drill bit, if the drill bit is not properly seated inside the screw or if you're not applying enough pressure while drilling. When you're securing the screws and you hear a clunk, clunk, clunk from the drill, that means that you're stripping the screw and you need to stop immediately. Put the drill in "reverse" to take the screw out just a bit, switch it to "forward" and drill it into the wood.

Replacing a Railing Board

Before beginning, remove all planter boxes, bird feeders, and anything else you may have attached to the deck railing.

Put on the safety glasses. The first step is to locate the fasteners (i.e., screws or nails) that connect the top rail to the posts or house, and remove them with a screwdriver or the claw of the hammer. Again, you may see that the builder just used whatever fastener was in his nail apron, so you may have a mixture of both.

The reality of working with wood is that the fasteners may be in too deep or are completely rusted, making them impossible to budge on the first, second, or third try. Since you're replacing the wood, don't be afraid to gouge the wood around the nail or screw with a Phillips screwdriver. The wood will tear away enough to provide you with better access to the fastener. If you're trying to remove screws, be careful not to pull on the board to loosen them because that may cause the face board to crack.

If you're having difficulty removing a nail, use a cat claw with the hammer. This works just about every time. If the screw isn't budging, you can use the Phillips screwdriver to gouge around the screw so that you can get more leverage. Don't worry about damaging the wood, because you're removing it, remember? So dig in, if you must!

And if all else fails, you can use a reciprocating saw. Wearing safety glasses, place the saw into the space between the railing board and the face board, turn the saw on, and cut through the fasteners.

If any nails are still in the board, hammer them down to prevent any injuries.

Place the board (or pieces) into a large garbage can.

Installing a New Railing Board

Place the railing board onto the posts. If you're replacing more than one board, you always want to be sure that the boards meet at the center of a post.

You'll be inserting 2 screws per every 6 inches of width and only where the board rests on the posts. And because you're using self-tapping deck screws, you won't need to create pilot holes.

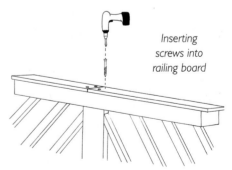

Inserting screws into railing board

Put on the safety glasses. Using the Phillips drill bit, insert the screws through the board and into the posts.

Note: You can strip a screw by using the wrong size drill bit, if the drill bit is not properly seated inside the screw or if you're not applying enough pressure while drilling. When you're securing the screws and you hear a clunk, clunk, clunk from the drill, that means that you're stripping the screw and you need to stop immediately. Put the drill in "reverse" to take the screw out just a bit, switch it to "forward" and drill it into the wood.

Tools Needed

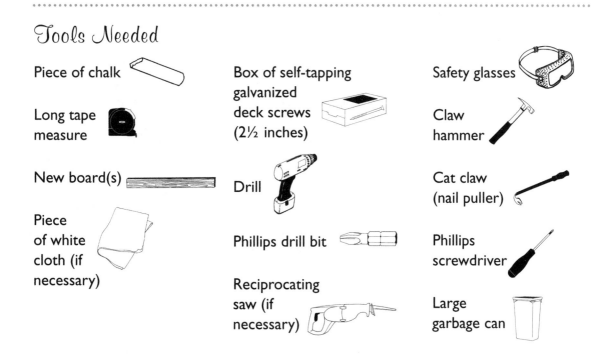

Piece of chalk

Long tape measure

New board(s)

Piece of white cloth (if necessary)

Box of self-tapping galvanized deck screws (2½ inches)

Drill

Phillips drill bit

Reciprocating saw (if necessary)

Safety glasses

Claw hammer

Cat claw (nail puller)

Phillips screwdriver

Large garbage can

Fixing a Wood Fence

*I*f the old adage **"Good fences make good neighbors"** is true, then what is your damaged fence saying about your relationship with the people next door? Maybe it's time to mend both.

A wood fence is composed of vertical posts that are cemented into the ground and separated typically about 8 feet (or less) apart. Then there are rails. These pieces of wood run horizontally between the posts, one connected at the top, one in the middle, and one at the bottom. The boards are the pieces that are attached vertically to the rails and cover the entire area of fencing.

You rarely have to mess with posts and rails. It's those darn boards that tend to be the problem because they bear the brunt of the weather . . . or deer.

If you see one board that's damaged, then there are probably a few more that you haven't noticed. Therefore, we recommend that you walk the interior perimeter of your fenced yard and carefully look at each board, top to bottom. Use a piece of chalk to mark ones that are loose or damaged. Then take count of how many boards need to be secured or replaced.

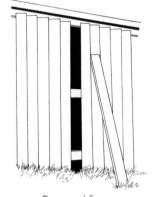

Damaged fence

Buying a Fence Board and Screws

You need to know the height and width of the board(s) you'll be replacing. Don't panic if the wood in the store is wider than the boards on your fence. When a board is cut at the lumberyard, it is cut to a specific size. Yet, by the time it arrives in a store, it has shrunk, and after you install it, it will shrink again. That's because the wood contains water, and the water evaporates over time.

If you're not going to cut the board yourself, then be sure to purchase the wood at a home improvement store or lumberyard that will cut the board for you.

The best fastener for securing a fence board is a self-tapping galvanized wood screw, typically 2½ inches. The self-tapping screws have a very sharp tip, which prevents the wood from splitting, kind of like drilling a pilot hole. The galvanized part means that the screw won't rust.

You'll need three screws per board, but it's best to just buy a box of them. And make sure that you have a Phillips drill bit for this project, too.

Fastening a Loose Board

If a board is loose (and by that we mean hanging off), chances are that it was secured with a nail instead of a wood screw. A nail can pop up, but a screw has threads that dig into the wood, making it difficult to move.

Wearing safety glasses, use the hammer to bang out the nail, hitting it on the point, not the head. Remove the nail. Repeat with the other nails in the board. Position the board in place. Using the drill with the Phillips drill bit, insert a screw into each hole.

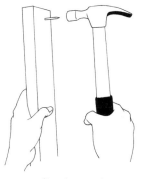

*Banging out the
point of a nail*

Note: You can strip a screw by using the wrong size drill bit, if the drill bit is not properly seated inside the screw or if you're not applying enough pressure while drilling. When you're securing the screws and you hear a clunk, clunk, clunk from the drill, that means that you're stripping the screw and you need to stop immediately. Put the drill in "reverse" to take the screw out just a bit, switch it to "forward" and drill it into the wood.

Replacing a Board

Position the new board into the opening in the fence and have your helpful friend hold it in place. Put on the safety glasses and have your friend wear a pair, too.

Using the Phillips drill bit, start at the top and insert a screw through the board and into the rail. Then insert a screw into the middle section, through the board and into the rail. And then insert a screw into the bottom, through the board and into the rail.

Inserting a screw

Note: You can strip a screw by using the wrong size drill bit, if the drill bit is not properly seated inside the screw or if you're not applying enough pressure while drilling. When you're securing the screws and you hear a clunk, clunk, clunk from the drill, that means that you're stripping the screw and you need to stop immediately. Put the drill in "reverse" to take the screw out just a bit, switch it to "forward" and drill it into the wood.

Tools Needed

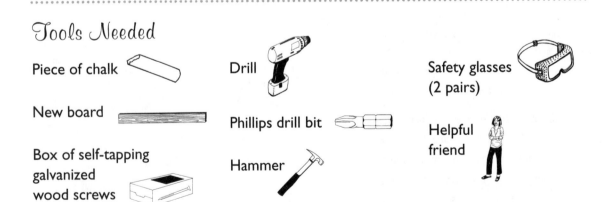

Piece of chalk

New board

Box of self-tapping galvanized wood screws

Drill

Phillips drill bit

Hammer

Safety glasses (2 pairs)

Helpful friend

Preventing a Wet Basement

 **o you run to the basement after every rainstorm to check for a possible flood? With these easy instructions you'll be singing in the rain . . . outside, that is.**

Two words a homeowner never wants to hear together are *wet* and *basement*. Those two words collectively mean only one thing: Trouble, *with a capital T.* The way to keep those words out of your vocabulary, and house, is to do some preventive maintenance.

Checking the Source of the Problem

Splash Blocks

The first thing to do is to walk around the exterior of your home to check that every downspout has its own splash block. Also, make sure that the splash block is located directly underneath the downspout and is positioned so that the water will flow *away* from the house. If necessary, you may need to pack some soil or sand underneath the narrow end of the splash block so that it sits up higher than the wider end.

If you need to purchase a splash block, we recommend that you buy plastic rather than concrete because it's lighter to carry and easier to clean.

Properly placed splash block

Gutters and Downspouts

When it comes to dealing with gutters and downspouts it's kind of like *out of sight, out of mind*, right? Since you can't easily see what's inside, you'd rather just think positive thoughts that they're perfectly clear of leaves, twigs, and birds' nests. And if not, then a heavy rainstorm will do the dirty work for you. We have two words that will snap you out of la-la land: *wet basement*.

A heavy rain, even days of torrential storms, will not clear out gutters and downspouts. In fact, if your home's drainage system is clogged, then you have a very good chance of getting water inside your home. To help keep that from happening, either clean the gutters and downspouts yourself (see our first *Dare to Repair* book), or pay to have it professionally done, at least twice a year, depending on where you live.

Cleaning gutter and downspout

Buying a Downspout Extender

Now that your home's drainage system has been cleared of debris and every downspout has a splash block, it's time to take one more step—installing a downspout extender.

There are different types of water diverting systems, ranging from a complicated underground system, which requires you to dig a long trench, lay rocks, and insert a long flexible plastic extender, to exposed extenders that can be easily mounted to a downspout in a matter of minutes. Installing an underground system can be a very big project, so try one of the exposed extenders first. Just remember that no matter which one you choose, you'll need it on all of your home's downspouts.

Installing an Exposed Downspout Extender

Note: For this project we used a jointed downspout extender because you can move it around to direct the water to different areas. Another plus is that you can lift the base to a 90-degree angle when you need to mow in that area. You just have to remember to place it down afterward.

Use the screwdriver to remove the screws on the elbow of the downspout. If the screws are rusted, spray them with WD-40 and wait about 5 minutes before trying again. Remove the elbow and put the screws in your pocket. Don't throw the elbow section out because you may need it for later use.

Downspout extenders can **be purchased online or in home improvement stores.**

Removing downspout elbow

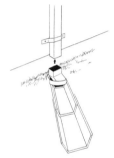

Attach downspout extender to downspout

Push the jointed downspout extender onto the bottom of the downspout. Most downspouts measure 3' × 4', but if your downspout is 2' × 3', simply remove the adapter on the extender (we recommend that you store this extra part for possible future use). Insert the screws into the holes and tighten with the screwdriver.

Tools Needed

Downspout extender

Small Phillips screwdriver

Flathead screwdriver (if necessary)

WD-40 (if necessary)

Sealing a Driveway

*J*udy prided herself on living in a home so cute and tiny, it was referred to as the "gingerbread house." But after many years of neglecting the small driveway, it turned from being eye candy to an eyesore. Judy thought it would be a piece of cake to get to find a contractor to do the job, but they all said it wasn't worth their time. So, Judy did some research and realized she could do the project herself by using a few tools and not too much money. Now it's home *sweet* home, again.

You saw your father do this and never gave it a second thought. And now, as a homeowner, you wish you had, right?

It's important to seal a driveway to protect it from the elements, as well as to keep it from cracking. The job may seem overwhelming, but it's really quite easy. And, the best part is that you should do it only every three years! The reason is that the sealant adheres best after thirty-six months or so.

Okay, what's the catch, you ask? Well, you need to apply two coats, with a four-hour waiting period in between. And this is a project where your schedule has to work around the project's needs, and not the other way around.

If the driveway is in bad shape (i.e., there are a lot of cracks, and tiny trees are growing in those cracks), you'll need to repair the driveway before sealing it. No matter how tempting it is to seal the driveway as is, it's important to prep it beforehand so that the cracks don't get bigger and the sealant will adhere (see our first *Dare to Repair* book). Therefore, you may want to do the repairs one weekend and seal the driveway the next.

Buying the Sealant and Tools

You need to know the size of the driveway so you'll get the right amount of materials—you don't want to start this project only to run out halfway through! Use a long tape measure to find the dimensions.

Driveway sealant comes in tubs, and they weigh a ton. Okay, maybe half a ton. So, you'll need to drive a car that has enough space to carry 4 or more tubs and is capable of supporting the extra weight. Also, find a flatbed cart at the store to make the sealant easier to load and unload.

Look on the tub to find the area that it will cover, as well as the warranty information. This is important because the better the warranty, the better the product. You'll pay more, but it's worth it.

Another very important thing is to get buckets with the same batch number for consistency in color—just as with paint. The information is listed on the label.

Getting Started

The most important factor for determining when to do this project is the weather. You need to do the repairs and sealing on warm, sunny days with no chance of rain during the following 24 to 48 hours. Follow the manufacturer's directions for the correct temperature and time needed.

The day before, use the edger along the driveway so that the grass won't interfere with the sealing.

The night before, tip over the *unopened* tubs of sealant. This will allow any sediment that's been sitting on the bottom to get properly mixed in. Leave the containers upside down until the next day.

Move any vehicles out of the garage and onto the street.

Use the leaf blower or push broom to clean off the entire driveway. Turn the spray nozzle of the garden hose to the most forceful setting and completely hose down the area. Fill a bucket with water and TSP-PF detergent, following the manufacturer's instructions. Use this along with a long stiff-bristled brush to clean the driveway.

The average driveway is two cars wide and four cars long.

Preparing the driveway

Sealing the Driveway

If the garage is at the end of the driveway, open the door(s) so that you can do some of the sealing while standing inside. Be sure not to lock your front door because you won't be able to access the house through the garage once you start.

Tip over the buckets so they're right side up. Use the utility knife to cut a piece of the lid so that you can pry it off. Don't throw out the lid, because you'll need to keep the container covered and in the shade as you're working.

Use a long paint stick to stir the sealant. Lay the stick on a plastic grocery bag.

The goal in applying sealant is to create a thin, even layer. The way to achieve this is to not overpour the sealant, and to start at either end of the driveway, not in the middle.

Standing in the middle of one end of the driveway, tip the bucket over and pour out the appropriate amount, about the size of the circumference of the bucket. With you and your helpful friend each armed with a squeegee, have your friend spread the sealant from the center out to her side of the driveway, while you spread the sealant from the center to your side. You'll find that it's like painting with a roller—it's that easy.

Stirring the sealant

Spreading the sealant

Repeat the process of stirring, pouring, and spreading all the way to the other end of the driveway. You'll want to work swiftly because, as you'll see, the sealant can dry quickly.

Once the sealing is complete, line up the empty containers (with lids) at the foot of the driveway, near the street, and string caution tape across to keep anyone from walking or driving onto it.

Blocking off driveway

Wait 4 hours, and then apply another coat.

Allow the sealant to dry for 24 to 48 hours, depending on the manufacturer's directions.

Cleaning Up

Toss out the used squeegee rollers into the garbage bag, but keep the handles. Use the wet rag to clean the tools. Wash your hands (and clothes, if necessary) with warm, soapy water to remove any sealant.

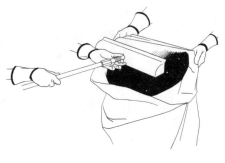

Cleaning up

Tools Needed

Long tape measure

Tubs of driveway sealant

Edger

Leaf blower or push broom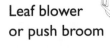

Garden hose with spray nozzle

TSP-PF detergent

Bucket

Stiff-bristled brush

Work clothes

Work gloves

Old shoes or boots

Utility knife

Long paint sticks

Plastic grocery bag

Helpful friend

Wide squeegees/ rollers (2)

Long handles (3—1 for the brush, 2 for the squeegees)

Caution tape

Large garbage bag

Wet rag

Installing a Flag Holder

Sixty-two percent of households in the United States fly an American flag. If yours is one of the 38 percent that doesn't, we know it's not because you're unpatriotic—you just didn't know how to install the holder. Well, with this simple project, you'll get your chance to show your true colors—red, white, and blue!

Buying a Flag with Flag Holder

Every American flag will have the same number of stars and stripes, but what will vary are the size, the material, and the holder.

Residential flags are made of poly/cotton, cotton, nylon, or spun polyester, with nylon being the most popular. The most common sizes for homes are 2.5' × 4' or 3' × 5'.

Flag holders are made of either plastic or metal. If the flag is big, you'll want to purchase a metal flag holder for better support.

Flags can be purchased separately, or as kits, which come with a flag holder and screws. These screws tend to be on the short side. Therefore, we recommend that you buy at least 1½-inch self-tapping galvanized wood screws and use those instead. Just be sure that the screws are not longer than the wood!

Getting Started

There's no rule in the U.S. Flag Code for where a flag has to be flown on a residence, so the decision is up to you. For ease of installation, the optimal choice is a wooden structure, such as a post on the front porch, or above the garage, or on the deck. If your home's facade is siding, brick, or stone, you're not out of luck. The framing around the

Caring for a U.S. flag: You can wash it with mild soapy water. Some dry cleaners offer to clean a U.S. flag for free.

windows and doors, including the garage, is wood. And even if that framing is covered with siding, you'll still be able to use it.

There's also no rule in the Flag Code for how high the flag should fly, so again it's up to you to decide the height for the installation. Just be sure to take into consideration any other flags (e.g., holiday motifs or college logos) you want to display with it, because the U.S. flag should *always* be flown higher than any other.

Installing the Flag Holder

Note: Before you begin, you want to be very certain of the location, especially if you'll be installing the flag holder into siding or brick. It's not like hanging a picture, where you get an easy redo—if you make a mistake or you change your mind about the site, you've caused damage to the structure. This is why we recommend that you use wood as the support for the flag holder.

If necessary, set up the ladder or stepladder, using the appropriate safety measures. Place the mounting bracket at the site where you will be installing the flag. Mark the holes (typically 4) with an awl and hammer, because a pencil may not fit through the holes.

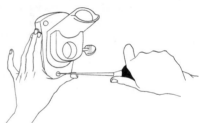

Marking mounting bracket holes

You may want to temporarily remove the wing nut and part of the bracket to make it easier to access the holes for drilling.

Position the mounting bracket over the holes.

Drill each self-tapping galvanized wood screw *almost* all the way in, and then go back and insert them completely into the wood.

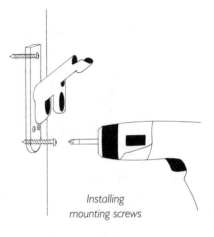

Installing mounting screws

Note: You can strip a screw by using the wrong size drill bit, if the drill bit is not properly seated inside the screw or if you're not applying enough pressure while drilling. When you're securing the screws and you hear a clunk, clunk, clunk from the drill, that means that you're stripping the

HERE IS A LIST OF SOME IMPORTANT HOLIDAYS ON WHICH TO FLY A U.S. FLAG

- New Year's Day
- Third Monday in January—Martin Luther King Jr. Day
- January 20—Inauguration Day
- February 12—Lincoln's Birthday
- Third Monday in February—Presidents' Day
- April 6—Army Day
- May 8—V-E Day
- Third Saturday in May—Armed Forces Day
- Last Monday in May—Memorial Day
- May 30—Traditional Memorial Day
- June 14—Flag Day
- July 4—Independence Day
- First Monday in September—Labor Day
- September 11—Patriot Day
- September 17—Constitution Day
- Second Monday in October—Columbus Day
- November 11—Veterans Day
- Fourth Thursday in November—Thanksgiving
- December 7—Pearl Harbor Day
- December 25—Christmas

screw and you need to stop immediately. Put the drill in "reverse" to take the screw out just a bit, switch it to "forward" and drill it into the wood.

Some mounting brackets come preset at a 45- or 90-degree angle. If it's adjustable, change it to the desired position.

Tools Needed

Flag

Flag holder assembly

Ladder or stepladder (if necessary)

Awl

Hammer

Self-tapping galvanized wood screws (1 ½ inch or longer)

Drill

Phillips drill bit

Resources

The following local, national, and international businesses have contributed their knowledge and/or products to *Dare to Repair, Replace & Renovate*. It gives us great pleasure to thank them publicly, and we hope that you'll contact them for information and products listed below.

Casablanca Fan Company
Dana Bigman
1-888-227-2178
www.casablancafanco.com

Ceramic Tile Education Foundation
Stephanie Samulski
www.tilecareer.com

The Container Store
1-800-786-7315
www.containerstore.com
(elfa closet system)

Cooper Lighting
www.cooperlighting.com
(security floodlight with indoor alarm module,
motion activated #MS249R)

Del Conca, USA

www.delconcausa.com

(wall and floor ceramic tiles)

Distinctive Renovations

1-703-898-8508

matt@distinctive-renovations.com

Ed Del Grande

Triple Master Contractor/Plumber

Author, *Ed Del Grande's House Call*

www.eddelgrande.com

Edmond's Paperhanging Co.

Edmond Fekecs

1-703-759-5514

Gardner-Gibson

1-800-237-1155

www.gardner-gibson.com

Great Falls Distinctive Interiors, Inc.

1-703-263-0180

www.gfdii.com

Home Design Elements

1-703-759-5800

www.hdelements.com

Home Safety Council

www.homesafetycouncil.org

Kohler
1-800-4-KOHLER
www.kohler.com
(K-169 kitchen sink faucet with K-171 sink sprayer)
(K-11000 Bol ceramic faucet)
(K-3654 Persuade™ dual flush toilet)

Long Fence
www.longfence.com

Lowe's
www.lowes.com

M-D Building Products, Inc.
1-800-348-3571
www.mdteam.com
(M-D's All Climate EPDM Rubber Weatherseal)

T-N-T Services Group, Inc.
1-301-874-8400
www.tntservicesgroup.com

Trunnell Electric
1-301-258-8300
www.trunnellelectric.com

Valley Forge Flag Company
Chris Binner
www.valleyforgeflag.com

Valspar
www.valspar.com
(paints)

Wallwik

www.wallwik.com

Werner Ladder Co.

www.wernerladder.com

Yards & Beyond

Brighten Your Outdoor Living!

www.jiaweisolar.com

(solar lights, premium grade #282818)

York Wallcoverings

LeRue Brown

1-800-375-YORK

www.yorkwall.com

Helping Others

Now that you've learned some new skills, why not put them to good use? Here are just a few of the wonderful nonprofit organizations that help build or renovate homes for the disadvantaged, while revitalizing neighborhoods as well.

Habitat for Humanity International
Women Build Program
121 Habitat Street
Americus, GA 31709
1-800-422-4828
www.habitat.org/wb

Habitat for Humanity ReStore
www.habitat.org/env/restores.aspx

The Fuller Center for Housing
701 S. Martin Luther King Jr. Blvd.
Americus, GA 31719
1-229-924-2900
www.fullercenter.org

Rebuilding Together
1536 16th Street, NW
Washington, DC 20036
1-800-473-4229
www.rebuildingtogether.org

Index